THE

SECRET OF PLATO'S ATLANTIS.

BY

LORD ARUNDELL OF WARDOUR,

AUTHOR OF

"TRADITION, PRINCIPALLY WITH REFERENCE TO MYTHOLOGY AND THE LAW OF
NATIONS;" "THE SCIENTIFIC VALUE OF TRADITION;" "THE NATURE
MYTH UNTENABLE FROM THE SCRIPTURAL POINT OF VIEW."

LONDON: BURNS AND OATES.

1885.

LONDON :
ROBSON AND SONS, PRINTERS, PANCRAS ROAD, N.W.

PREFACE.

THE following pages were written for the *Month*, but in the course of writing extended themselves beyond the limits of a magazine article; the third chapter more particularly becoming too elaborate in form for suitable publication in a periodical. I have, therefore, preferred to publish them separately. As, however, it would have involved too much trouble to have rewritten and recast the articles, I have printed them in their original form, as addressed to the readers of the *Month*.

The subject, at least, is a curious and interesting one; and Mr. Donnelly's work, which was the occasion of the articles being written, contains much curious speculation, and is written in a style calculated to give zest to the inquiry. It has had a wide circulation.

I cannot expect the same circulation for this little volume, more especially as the theory it offers is not of the same romantic and popular character; but I hope it may contribute something towards the solution of an interesting and difficult question.

CONTENTS.

PLATO'S ATLANTIS.

CHAPTER I.

PLATO'S ATLANTIS—MR. DONNELLY'S THEORY.

A BOOK which is now (1883) in its seventh edition seems to claim some reply from the point of view of Tradition. It is entitled *Atlantis': the Antediluvian World*,* and, in fact, announces that the Deluge, in which we have hitherto believed and have called universal, at any rate to the extent of the destruction of all mankind,† did not really occur, but that the subsidence of the island or continent of Atlantis at some indefinite period was attended by very similar circumstances, and that it is the tradition of this catastrophe which has somehow spread through all countries, which has created the impression of a universal deluge ; in other words, that there was a deluge, but a deluge as revealed according to Plato, and not according to Moses.

The evidence which Mr. Donnelly has accumulated, both as to the diluvian tradition and also as to the common

* *Atlantis : the Antediluvian World.* By Ignatius Donnelly. 7th edition. (Sampson Low) London, 1883.

† This chapter was written previously to the controversy on the Deluge in the pages of the *Tablet* in the year 1884. I am not, however, aware that anything transpired in that controversy which would require me to retract or modify any statement in the present paper. If so, I shall be obliged to any one who will put his finger on it.

origin, at any rate, of the civilised nations " on both sides
of the Atlantic," is by no means inconsiderable ; and it
will be seen that, in so far as he fails to sustain his special theory of the submerged Atlantis, his convictions,
facts, and testimonies must pass to the account or lapse
to the inheritance of what I have regarded as the tradition
of the human race.

As it is always safer and fairer to present the theory
of an author in his own words so far as may be possible,
I will give the principal heads under which Mr. Donnelly
summarises the purpose of his work.   I shall have occasion, at any rate indirectly, to refer to the omitted headings
in the course of this discussion :

" 1. That there once existed in the Atlantic Ocean, opposite
the mouth of the Mediterranean Sea, a large island, which was
the remnant of an Atlantic continent, and known to the ancient
world as Atlantis.   2. That the description of this island given by
Plato is not, as has been long supposed, fable, but veritable history.
3. That Atlantis was the region where man first rose from a state
of barbarism to civilisation.   4. That it became in the course of
ages a populous and mighty nation, from whose overflowings the
shores of the Gulf of Mexico, the Mississippi, the Amazon, the Paci-
fic coast of South America, the west coast of Europe and Africa, the
Baltic, the Black Sea, and the Caspian were populated by civilised
nations.   5. That it was the true antediluvian world—the Garden
of Eden, the Garden of the Hesperides, the Elysian Fields, the Garden of Alcinous, the Mesomphalos, the Olympos, the Asgard of the
traditions of ancient nations ; representing a universal memory of
a great land where early mankind dwelt for ages in peace and
happiness. . . . 12. That Atlantis perished in a terrible convulsion
of nature, in which the whole island sank into the ocean, and nearly
all its inhabitants.   13. That a few persons escaped in ships and
on rafts, and carried to the nations east and west the tidings of
the appalling catastrophe, which has survived to our own time in
the Flood and Deluge legends of the different nations of the old
and the new worlds."

In this theory there are two distinct propositions :
(1) that an island or continent of Atlantis existed, and

sank in the ocean; (2) and that this submersion was the origin of the various diluvian legends which are found in all parts of the world.

The legend of Atlantis can hardly be asserted even by Mr. Donnelly to be the tradition of the human race, for he himself terms it "a novel proposition."

" The fact that the story of Atlantis was for thousands of years regarded as a fable proves nothing. There is an unbelief which grows out of ignorance as well as a scepticism which is born of intelligence. . . . For a thousand years it was believed that the legends of the buried cities of Pompeii and Herculaneum were myths. . There was a time when the expedition sent out by Necho to circumnavigate Africa was doubted, because the explorers stated that, after they had progressed a certain distance, the sun was north of them. This circumstance, which then aroused suspicion, now proves to us that the Egyptian navigators had really passed the equator, and anticipated by 2100 years Vasquez da Gama in his discovery of the Cape of Good Hope " (p. 3).

On the other hand, although it does not appear that Mr. Donnelly himself believes in the inspiration of Genesis, yet the fact that it has been so believed by many millions in many parts of the world during a long continuance of years must stand for something as against a theory.

As it is my wish to confine my argument to the limits of historical tradition, I should have been willing to have accepted Mr. Donnelly's first proposition, viz. that Atlantis existed and subsided, at any rate *pro argumento*, if historical investigation had not destroyed the *primâ facie* evidence which seemed to compel or invite the inquiry. This, however, is a point which the reader must decide. Apart, however, from the historical evidence, I must remark that Mr. Donnelly's theory is opposed, from their several points of view, by Mr. Wallace, Mr. Darwin, and Professor Geikie (*vide* Wallace's *Island Life*, chap. vi. 11).

Mr. Wallace's argument is not, it is true, addressed to

the same set of facts as are adduced in Mr. Donnelly's chaps. v. vi.—"The Testimony of the Sea," "The Testimony of the Flora and Fauna." This, however, is a matter which Mr. Donnelly must settle with Mr. Wallace. The date of Mr. Donnelly's first edition is not stated.

Mr. Donnelly's second proposition is, of course, dependent on the first; but I will continue the analysis of his evidence. If the existence of Atlantis could have been considered probable, we might have believed it to have been the scene of the earthly paradise, the location and domicile of man in the antediluvian world, and the direction to which alike the sad reminiscences and bright hopes of mankind reverted.

I will now proceed to discuss the principal evidence which Mr. Donnelly adduces. There is one testimony at p. 95 which seems in some sort to favour the suggestion I have just made: "The traditions of the early Christian ages touching the Deluge pointed to the quarter of the world in which Atlantis was situated." This, however, is only based on the theory of the good monk Cosmas, who believed that the world was flat. "There was a quaint old monk named Cosmos [Cosmas] who, about a thousand years ago, published a book, *Topographia Christiana*, accompanied by a map [an engraving of which is given], in which he gives his view of the world as it *was then understood.**** It was a body surrounded by water, and resting on nothing. . . . It will be observed that while he locates Paradise in the East, he places the scene of the Deluge in the West, and he supposes that Noah came from the scene of the Deluge to Europe." In Dr. Smith's Greek and Roman biography it is, however, said on the contrary: "The object of this treatise is to show, in opposition to the universal opinion of astronomers, that the earth is not spherical, but an extended surface. . . . .

---

* Italics throughout are mine, unless the contrary stated.

Weapons of every kind are employed against the *prevailing theory*," &c. And although he quotes *inter alia* the authority of the Fathers, it will hardly be disputed that the prevailing Christian opinion, commencing with Gen. xi. 2, "And when they removed from the East" to the plain of Sennaar, has located the descent from the Ark in the mountains of Armenia.

"We have already seen that Berosus relates how in his time portions of the Ark were removed and used as amulets. Josephus says that remains of the Ark were to be seen in his day upon Ararat. Nicolas of Damascus reports the same. St. Epiphanius writes, 'The wood of the Ark of Noah is shown to this day in the Kardæan [Koord] country'" (*Adv. Hæres.* lib. i.; *Legends of Old Testament Characters*, S. Baring-Gould, i. 165).

So much, at any rate, as to the prevailing opinion. Cosmas, before he had become a monk, had been a great navigator, but his explorations had been in the Indian Ocean.

Mr. Donnelly is necessarily limited to the data found in the fragment of Plato. Plato commences with this statement :

"The tale, which was of great length, began as follows. I have before remarked, in speaking of the allotments of the gods, that they distributed the whole earth into portions differing in extent, and made themselves temples and sacrifices. And Poseidon, receiving for his lot the island of Atlantis, begat children by a mortal woman. . . . He also begat and brought up five pairs of male children : . . . the eldest, who was king, he named Atlas, and from him the whole island and the ocean received the name of Atlantis" (p. 13).

Now, as to Poseidon, I recommend Mr. Donnelly to a short but able treatise—*Poseidôn : a Link between Semite, Hamite, and Aryan.* By R. Brown (Longmans, 1872)— in which the worship is traced " from its starting-point in Chaldæa, through Phœnicia, Philistia, Libya, and Greece;" and Mr. Brown finally identifies him with the patriarch

Noah, as handed down in Libyan mythology, following in this the lines of tradition.*

And looking to the diffusion of this worship of Poseidon in Africa, including Egypt, Carthage, Ethiopia, Mauritania, and throughout the Phœnician colonisation, we seem to understand Plato's statement that Atlantis once "had an extent greater than Libya and Asia." "For many centuries," says Lenormant, "the Pelasgi of the Archipelago, Greece, and Italy, the Philistines of Crete, the Sicilians, the Sardinians, the Libyans, the Maxyans of Africa, in spite of the distance of sea separating them, united in a close confederation, maintaining a constant intercourse, and thus explaining the Libyan element, hitherto inexplicable, in the most ancient religious traditions of Greece, the worship of the Athenian Tritonis and of the Libyan Poseidon."

Atlantis takes its name from Atlas—"the king." We hear of Atlas first in Hesiod, as son of Japetus; his brother was Menœtius ("Mnesius," Plato; "Menu," Lenormant), and, according to Apollodorus, his mother's name was *Asia*. In the Homeric poems he knows all the depths of the sea; he bears the long columns which tear asunder or carry all around earth and heaven: in either case the meaning of keeping asunder is implied. Atlas is also described as the leader of the Titans in their

* Mr. Brown's argument would have been much enforced if he had noticed the following passage in the *Journal of the Asiatic Society*, xv. p. 231, by Colonel Rawlinson, C.B.: "I read the two names—the cuneiform writing cannot be transferred to your columns—doubtfully as Sisiron and Nahu (Nosh) That the god in question represents the Greek Neptune is, at any rate, almost certain; he was worshipped on the *seashore* and ships of gold were dedicated to him. His ordinary title . . . and the latter word is explained in the vocabulary as . . . that is, ' apzu,' which may be allied to ' Ποσ ' in Ποσειδων, as it is also joined with ' nun,' a fish. His other epithets are . . . ' sur marrat,' ' king of the sea,' and . . . probably ' god of the ship or ark.' Other titles I cannot explain; but they seem to be all connected with traditions of the biblical Noah."

contest with Zeus; others represent Atlas as a powerful king, who possessed great knowledge of the course of the stars (Smith's *Dictionary*).

In the Targums, Nimrod is thus made to address his subjects: "Come, let us build a great city. . . . In the midst of our city let us build a high tower. . . . Yea, let us go further; let us *prop up the heaven* on *all sides* from *the top of the tower*, that it may not again *fall* and *inundate* us. Then let us *climb up* to heaven and break it up with axes. . . ." (Baring-Gould, *Old Testament Characters*, i. 166).

We may be allowed to conjecture, then, that either Atlas is the tradition of Nimrod, or Nimrod of Atlas. Will Mr. Donnelly maintain the latter in face of the historical evidence of Nimrod in the Bible, and in the cuneiform tablets?* Among other sons of Poseidon who bear

* In the *Month*, January 1884, I discussed the evidence as to the historical existence of Nimrod with reference to the cuneiform tablets. It has struck me since that the direct evidence, so far as I know, has never been collated with the indirect evidence, as, for instance, as to the existence of Chus, the father of Nimrod. Now, for this there is the testimony not only of Asia, but of Africa. As regards the latter, there is the testimony of Josephus, recording the Gentile evidence of his day, and the independent recent evidence of the Egyptian monuments. Josephus says (*Ant.* i. vi. 2): "Some indeed of its names (descent of Ham) are utterly vanished; . . . yet . . . time has not hurt at all the name of 'Chus;' for the Ethiopians, over whom he reigned, are even at this day, both by *themselves* and by all men *in Asia*, called *Chusites*. The memory also of," &c. That this testimony of Josephus is corroborated by the most recent evidence will be apparent from the following references to Brugsch's *Egypt* (i. 284): "We have substituted for the *Egyptian* appellations Ta-Khont and *Kush*, the better known names Nubia and Ethiopia;" (ii. 76) "the land of Kush;" and upon the Assyrian conquest of Egypt, B.C. 1000, we find the name Nimrod reappearing (ii. 206): "for Takeloth, Usarkon, Nemaroth represent in the Egyptian form writing the names Tiglath, Sargon, and Nimrod, so well known in Assyria." As regards Asia, the tradition had been fully recognised (*vide J. of Asiatic Soc*, v. xv. pp. 230-33): "In Susiana the chief seat of the *Cush*, we have the Scythic "Scythic or Hamitic," [p. 232] inscriptions of Susa and Elymais, and the Scythic names of Kissia, Cossica, Shus Afar, &c., not forgetting the tradition of the Ethiopian Memnon and the Ethiopian Cepheus. Along the line to India the Ethiopians

resemblance to Atlas and Nimrod are Orion, " the giant
hunter " (" Nimrod is called in the LXX. the giant
hunter,"), and " the colossal youths Ôtos and Ephialtes,
who at nine years old attempted to scale heaven by piling
up mountains; which, says Homer, they would have
accomplished had not Apollo slain them. . . . Mr.
Gladstone remarks that the efforts of the two youths
recall the traditions of the Tower of Babel " (*Juv. Mun.*
251 ; Brown's *Poseidon*, 84).

Mr. Donnelly's best point is his suggestion that Atlantis is identical with Aztlan in Central America :

" Upon that part of the African continent nearest to the site of
Atlantis we find a chain of mountains known from the most
ancient times as the Atlas mountains. Whence the name Atlas,
if it be not from the name of the great king of Atlantis ? . . .
Look at it ! An Atlas mountain on the shore of Africa ; an Atlan
town on the shore of America ; the Atlantis living along the north-
west coast of Africa ; an Aztec people from Aztlan in Central
America ; an ocean rolling between the two worlds called the
Atlantic ; a mythological deity called Atlas holding the world on
his shoulders ; and an immemorial tradition of an island of Atlantis.
Can all these things be result of accident ? " (p. 172.)

We shall presently have to consider the question how
far the " immemorial tradition " is the offspring of the
invention of Plato. Before abandoning the present ground,
let me remark that one form of the legend of Atlas makes
him King of Mauritania, where are also located the
mountains of Atlas and the Atlantis. Atlas was fabled
to have been turned into a mountain by Perseus, who was
refused hospitality by Atlas, because he had been informed

---

of Southern Persia were known to Homer, Herodotus, and Strabo. . . .
The Brahni division of the Belûs rejoined their Cushite brethren in
Mekran by crossing from Arabia, and still speak a Scythic dialect;
while the names of Kooch and Belooch for Kus and Belûs remain to the
present day."—Colonel Rawlinson, C.B. (now Sir Henry Rawlinson,
K.C.B.).

by an oracle of Themis that he should be dethroned by
one of the descendants of Jupiter. This reads very much
like the tradition that the descendants of Japhet were to
dwell in the tents of Canaan; and the belief of Atlas
having been subdued by Perseus the Grecian hero—the
friend of Athene—may account for that part of the speech
put into the mouths of the Egyptian priests by Plato:
" Many great and wonderful deeds are recorded of your
State in our histories; but one of them exceeds all the
rest in greatness and valour; for these histories tell of a
mighty power which was aggressing wantonly against the
whole of Europe and Asia, and to which your city put an
end. This power came forth out of the Atlantic Ocean,"
&c. However, I shall give later on an alternative suggestion.

The inadvertent reader needs to be very much on his
guard in reading Mr. Donnelly. Each subsequent chapter
absolutely assumes the conclusions of the previous chapter. Thus, ch. vii., " The Irish Colonies from Atlantis,"
which naturally excites our interest, commences, " We
have seen that beyond question Spain and France owed a
great part of their population to Atlantis;" but if we
revert to ch. iv., " The Iberian Colonies of Atlantis,"
with the exception of the statement, which I shall presently discuss, that the Turdetani are said by Strabo to
have had writings 6000 years old, there is nothing whatever tending to support his contention. There is, indeed,
the assertion that the Basque language has analogies
with the Algonquin and other American languages; and
there is a similar argument in another very learned
chapter in respect to the affinity between the Maya and
Phœnician. I remember reading in the *Month* that
the devil is said to have spent two years in the Basque
country endeavouring to learn the language, but at the
end of that time abandoned it, as he had only mastered

one word, which was written like "Nabuchodonosor" and
pronounced "Sennacherib." Allowing, however, Mr.
Donnelly to have seen farther into the millstone than any
one else, this correspondence of language would tend to
prove the common origin of mankind, the original unity
of tongue, and the migration from a common centre in
Mesopotamia equally with emigration from Atlantis;
unless, indeed, the reader is prepared to believe that the
"Mayas" of America are descended from "Maia," the
daughter of Atlas! The Iberians having been thus
demonstrated to be "Atlanteans," it suffices to show in
the chapter on Ireland that the early invasions came from
Iberia. "Spain in that day was the land of the Iberians,
the Basques, that is to say of the Atlanteans" (p. 409).
Again we read (p. 286):

"We find the barbarians of the coast of the Mediterranean regarding the civilised *people of Atlantis* with awe and wonder. 'Their
physical strength was extraordinary, the earth shaking sometimes
under their tread. Whatever they did was done speedily. They
moved through space almost without the loss of a moment of time.'
This probably alluded to the rapid motion of their sailing vessels.
'They were wise, and communicated their wisdom to men.' That
is to say, they civilised the people they came in contact with."

Other quotations follow, all with reference to Murray's
Mythology.

We should naturally expect that these quotations from
Murray had some reference to Atlantis. Not at all. Mr.
Murray is only speaking of the Olympians. But Mr.
Donnelly having satisfied himself that Olympos is identical
with Atlantis (he even contends that the letters of the
words are interchangeable and the names identical), henceforward everything that is recorded of Olympos is con-
vertibly to be spoken of Atlantis.

From one point of view, Atlantis and Olympos, Asgard
and Atlantis, are part of a common tradition, a question
which I shall presently discuss. When, however, Mr.

Donnelly recognises resemblances, they must at once be regarded as conclusive, *e.g.* that Olympos is a tradition of Atlantis. In short, Mr. Donnelly appears uniformly to argue according to the formula " Cæsar and Pompey very much alike, especially Pompey."

It seems unnecessary to say that Mr. Donnelly sees the name of Atlantis everywhere. Except when he clutches at evidence in this way he appears perfectly able to weigh facts and evidence ; and it must be acknowledged that there is a seeming confirmation of his theory in the my- thological and classical location of the Garden of the Hesperides in the Islands of the West. I have already (p. 2) quoted Mr. Donnelly on this head. His confirma- tion of the theory, however, again disappears when we remember that the Garden of the Hesperides was only one of the reminiscences of Eden. It is true that from his point of view Eden is only a reminiscence of Atlantis ; but apart from the argument which I shall proceed to put— Eden in the East having been the prominent belief of mankind—the *onus probandi* lies on his side of showing that all those traditions, Meru, Olympos, Elysium, Asgard, Midgard, centred in Atlantis. So far from this being the case, the salient features of the tradition which are common to the other legends are barely discernible in the descrip- tion of Plato, *e.g.* instead of a garden we have only a fertile plain. With the exception of the Garden of the Hesperides all these other traditions place the Garden of Paradise in the East,* or the supposed centre of the world.

In all these legends (we shall agree so far) we find the embodiment of early tradition in a garden or a plain, a palace on or in connection with a mountain. There is,

* Moreover, the Bible and the Babylonian tradition placed paradise and the "father of countries " in the East (*vide* M. Lenormant, M. Oppert, and L'Abbé Vigouroux, *La Bible et les Découvertes Modernes*, i. 196).

however, one feature common to them all which, at first sight, favours Mr. Donnelly's theory, and which, perhaps, has confirmed him in it—they are all surrounded by water. This he naturally contends means the island of Atlantis. But when we consider that whenever the Ancients represented the *world* they represented it surrounded by water —it is so represented in Homer and in the map which Mr. Donnelly gives of the old monk Cosmas ; that one form the legend takes is that of Midgard, the middle of the earth, the "mesomphalos," which "was equally distant on all sides from the sea ;" and when we consider that according to the experience of mankind in their explorations in three directions, in the Atlantic and round the African coast to the Chinese seas, all was water—the north being sealed to them, as it is to us—I think Mr. Donnelly has only to enlarge his view, and he will fall back into the tradition of mankind.

At p. 326, Mr. Donnelly says : " Thus the nations on the west of the Atlantic look to the east for their place of origin ; while on the east of the Atlantic they look to the west : thus all the lines of tradition converge upon Atlantis." But precisely the same may be said if we start mankind from the plain of Sennaar.

And if we start mankind from the plain of Sennaar on the lines of the biblical narrative, is it unnatural to expect that they should embody their traditions of paradise, the Tower of Babel, and the Deluge in the conception, grotesque no doubt, of a garden on a mountain surrounded by water ? " In all the legends of India the original seat of mankind is placed on Mount Meru, the residence of the gods, a column uniting heaven to earth" (Lenormant, *Frag. Cosmog. de Berose*, p. 300). In the Scandinavian legend, " the centrical fortress which the gods constructed from the eyebrows of Ymen, and which towered from the midst of the earth equally distant on all sides from the sea, is certainly the Meru of the Hindoos and Indo-Scythæ. . . . It was the peculiar residence of the hero-god immediately after the Deluge ; and it is at once described with all the characteristics of a paradise, and is represented as a fortress which might secure the deities against any further attacks of the giants " (S. Faber, O.P.I., i. 220). " According to this creed " (the mythology of the Eddas) " Æsir and Odin had their abode in Asgard, a lofty hill in the centre of the habitable earth in the midst of Midgard—that middle earth which we hear of in early English poetry, the abode of gods and men. Round that earth, which was fenced in against the attacks of ancient and inveterate foes by a natural fortification of hills, flowed the great sea in a ring, and beyond that sea was Utgard, the outlying world, the abode of frost and giants and monsters, those old natural powers who had been dispossessed by Odin and the Æsir when the new order of the universe arose " (G. Webbe Dasent, *Tales from the Norse*, lvii.). Lenormant (301-2) says that the Iranian tradition corresponds. Hierapolis, Delos, and Ecbatana were constructed with reference to this tradition, and I consider that I have proved that the ancient state of Meroe, in the island of Meroe, near Mount Gibbainy—in the country of the Soudan (*vide Scientific Value of Tradition*, pp. 161 to 179)—was organised with reference to this tradition. The tradition of paradise in connection with the Deluge and the Tower of Babel is also seen in the hanging gardens (the paradisiacal mountain), the Pyramids in stages, and the tower of Borsippa, near Babylon (*vide* Leuormant, *id.* 318 *et seq.*). There is special mention (p. 320) of a bas-relief in the palace of Assur-bani-pal, of which a fragment has been published in Rawlinson's *Five Great Monarchies*, i. 388, " where a royal paradise adjoins a palace planted with large trees placed on the summit of an eminence, and watered by a single stream of

water, which divided itself into several channels on the
side of the mountain, like the stream of paradise, the
spring of Arvanda or Ardavâ-curâ of the Iranian Hara
Berazuiti, and the Ganga of the Indian Meru " (Lenormant,
321). Compare this with Plato's description of Atlantis :

" On the side towards the sea, and in the *centre* of the whole
island, there was a plain which is said to have been the fairest of all
plains, and very fertile. Near the plain again, and *also in the centre*
of the island, at a distance of about fifty stadia, there was a mountain not very high on any side. In this mountain there dwelt one of
the earth-born primeval men of that country whose name was Evenor . . . and Lucippe, and they had an only daughter named Chito.
. Poseidon fell in love with her . . . and breaking the ground
enclosed the hill in which she dwelt all round, making alternate
zones of sea and land, larger and smaller, *encircling* one another ;
there were two of land and three of water, which he turned as with a
lathe out of *the centre* of the island, *equidistant every way*, so that no
man could get to the island, for ships' voyages were not yet heard of.
He himself, as he was a god, found no difficulty in making special
arrangements for the *centre* island, bringing two streams of water
under the earth, which he caused to ascend as springs, one of
warm water and the other of cold, and making every variety of food
to spring up abundantly in the earth. He also begat and brought
up five pairs of male children, dividing the island of Atlantis into
ten portions " (Donnelly's *Atlantis*, p. 13).

I have given these extracts in juxtaposition at some
length, as it will thus be possible to decide whether those
including Atlantis are all common traditions of the one
historic narrative which embraces and completes them all,
or whether they all developed out of the slender reminiscences recorded of Atlantis.

I assume that Mr. Donnelly will intrench himself in the
position, as it seems to me the only position that remains
to him, viz. that

" Plato states that the Egyptians told Solon that the destruction
of Atlantis occurred nine thousand years before that date, to wit,
about nine thousand six hundred years before the Christian era.
This looks like an extraordinarily long period of time, but it must be

remembered that geologists claim that the remains of man found in the caves of Europe date back five hundred thousand years " (p. 29).

So tremendous a position can only be taken by a process of sapping and mining, as I confine myself to the historical facts, and do not profess to have at command such heavy artillery as will discharge 500,000 years in a single explosion.

Considering that all chronologies and histories upon analysis seem to terminate about 3000 b.c.—or, if they include the antediluvian world, to about 6000 b.c. ; that if the migration of the nations is retraced they are found to converge upon the central district lying between Persia and the Mediterranean, Armenia to Ethiopia ;* that, according to Mr. Proctor, the constellations known by similar names to variously dispersed nations can astronomically be shown to have been so named within the latitudes indicated above and about the year 2200 b.c., there is a background of probability for traditions tracing back to that period ; and, as against Mr. Donnelly, the argument might almost be stated mathematically. Given the amount of scepticism which will attach to the transmission of traditions of such calibre as the garden of paradise, the universal Deluge, the dispersion during 3000 years, how much will exist as to the preservation of the slight reminiscences of Atlantis, as above, during 9000 years ?

As regards the reminiscences of Atlantis, either the tradition of this palace, mountain, and canals was pre-

* For one instance, take what Colonel Rawlinson (sup. p. 232) says of the migration of the Scyths or Hamites : "They must have spread themselves at the same time over Syria and Asia Minor, sending out colonies from one country to Mauritania, Sicily, and Iberia, from the other. . . . It is well known to ethnographers that the passage of the Scyths is to be traced along all these lines, either by direct historical tradition, or by the cognate dialects spoken by their descendants at the present day. . . . And if we were to be thus guided by the mere intersection of linguistic paths, and independently of all reference to the scriptural record, we should still be led to fix on the plains of Shinar as the focus from which the various lines had radiated."

served before or after the subsidence of Atlantis. If *before*, how explain the fact that this tradition so curiously runs into the lines of the diluvian tradition ? They must, then, have been traditional of an event which happened *ex hypothesi* at later date ;. or if after, how explain that what would then be the direct tradition of the Deluge, or submersion, was thus transmitted only in an indirect, disguised, and legendary form ; and, on the other hand, that an apparently direct record of it, as in Genesis, should, in fact, be only the tradition at secondhand of a tradition in indirect form ?

The biblical record, the cuneiform narrative, the Indian legend, &c., all profess to give the tradition in direct form. How is it that they all tell of a universal deluge, in which one family—sometimes one man—survived, and that in all the prominent cause of the destruction was unintermittent and protracted rain ? In the case of Atlantis the cause was subsidence, or else the geological argument must be abandoned. Moreover, if the intelligence of the calamity, which was ultimately to take the form of the diluvian tradition, was to be extended piecemeal over the whole human race even in 9000 years, it could scarcely have been through one man or one family, but through many ; and it would seem none of the records or traditions tell of the event in the manner it is supposed to have happened, either according to the geological evidence, or according to the revelation of Plato.

In *Tradition* and elsewhere I have endeavoured to collate, though very imperfectly, the various traditions of the patriarch Noah in Chronos, Poseidon, Saturn, Hoa, &c. Chronos no doubt was the father of Poseidon, and so on ; yet fundamentally, whilst accreting other traditions, as of Shem, Cham, and Japhet, they all described a primeval legislator, who inaugurated or appeared in connection with a new order of things ; all came out of or had relations with, without being identified with, the ocean ; and although all are associated with the recollection of a primeval paradise, a golden age, a period of happiness and prosperity which was lost to mankind, they are almost all associated in some way with a catastrophe or with calamity. | They all plant the vine or the olive, so that it has been said "that all nations have given the honour of the discovery of agriculture to their *first* sovereigns." Now it happens, for the purposes of his argument, to be convenient to Mr. Donnelly to recognise this in part, and to apply it in this way. According to the requirements of his theory, the intelligence of the submersion of Atlantis was conveyed by the survivors to the various nations. He skilfully seizes hold of the tradition to which I have just referred, in order to despatch the various legendary heroes, no longer as representatives of the patriarch Noah, but, so to speak, on their own account to the various nations as the survivors of the catastrophe, and as the civilisers and legislators of the countries to which they came. Thus Hoa, or Hea, is despatched by him to Assyria (p. 83) : "He it was who was said to have brought civilisation and letters to the ancestors of the Assyrians. He clearly represented an ancient maritime civilised nation ; he came from the ocean, and was associated with some land and people that had been destroyed by rain and inundations." In like manner Saturn is sent to Latium,; but although the tradition is connected also with Kronos and Poseidon, and although it is said (p. 82) that "Chronos and Saturn were the same," yet Kronos and Poseidon are not so distributed, for the obvious reason that they stand at the commencement of the civilisation of Atlantis ! But this affords a measure for testing the theory. If "Chronos and Saturn are the same," Chronos cannot both be the father of Poseidon, who is gravely regarded by Mr. Donnelly as the

C

actual founder of the kingdom and dynasty of Atlantis, and at the same time the survivor after its subsidence (which happened after the lapse of "many generations") who brought civilisation to Latium.

The tradition of Saturn in Latium, I admit, fits in very well with Mr. Donnelly's theory, better even than he seems aware. I should like, however, to know where Mr. Donnelly finds mention "of 'a great Saturnian continent' in the Atlantic Ocean"? Mr. Donnelly is not lavish of references, and, until he gives one in this instance, I can only surmise that it is a free Transatlantic translation of the "Saturnia regna" of Virgil.

It may be, as Mr. Donnelly believes, that "Chronos and Saturn are the same," and yet that they represent the tradition at different stages and dates, and in Latium at the later date.

Sanchoniathon (*apud* Eusebius) says "that Chronos and Il are the same," and Lenormant says the same of Chronos and the Chaldean Ilu. Here we have the tradition at its earliest stage, and it will be worth while giving an extract from M. Lenormant, as it shows close resemblance with the tradition of Chronos, through Poseidon, in Plato's Atlantis:

"Ilu, the supreme mysterious god whom the Greeks have constantly likened to their Kronos. . . . The part which tradition, as recorded by Berosus, makes him play in the deluge is not perhaps without reference to one of his ideographic names; . . . for the complete group certainly reads Ilu—for example, in the name of Babylon—Bab-ilu; the sign . . . of which the primitive hieroglyph which we possess in some monuments represents a land intersected (*coupée*) by canals, is explained in the syllabaries by the root . . . which in Hebrew signifies 'to cleanse,' and in Assyrian 'to inundate.' It is thus 'the god of inundation, the god of the deluge'" (viz. Oppert, *Expédit. en Mésopotamie*, ii. 67; Lenormant, *Frag. de Berose*, 288).

We have seen Poseidon in Atlantis encircling his hill

with alternate zones of sea and land, and in the description of his palace the canals which he constructed are twice referred to. If these are common diluvian traditions of Kronos and Poseidon, it must follow that in Plato's account of Atlantis we may have diluvian traditions before the alleged period of its subsidence, *quod est impossibile*. *Ergo*, I should infer, a conclusion at which I shall arrive more definitely by another route, that Atlantis was, in the main, only general tradition taking form and embodiment in the mind of Plato.

I have still to notice the single fact upon which rests the foundation of chaps. iv. vii., that the Iberians, Gauls, and Celtic-Irish were Atlantes; viz. that Strabo tells us that " the Turdetani had written books containing memorials of ancient times, and also poems and laws set in verse, for which they claim an antiquity of six thousand years." Unfortunately, if we are to argue on Mr. Donnelly's lines, and if the submersion of Atlantis took place 9000 B.C., writings extending back only six thousand years do not help us at all. It is singular, however, that this figure should have been named by Strabo as dating *anno mundi* —6000 would be very nearly the correct date in his time. Mr. W. Palmer in his synchronism, " within five years four months and seven days," of the Hebrew and LXX., with Josephus and the Egyptian Chronicle, makes the commencement of the world *circa* B.C. 5360. I may add that, so long as Mr. W. Palmer's system remains unrefuted, we may be entitled, at any rate, to prefer his conclusions to the assertions of the Egyptian priests confuted by the testimony of their own monuments.

In Plato's description of Atlantis prominence is naturally given to the *horse*, as is appropriate in any mythological legend which commences with Poseidon. Mr. Brown (*Poseidôn*, p. 64) and also Mr. Gladstone (*Juventus Mundi*) are much exercised by this " remarkable connection of Poseidon with the horse." I am now only concerned with the fact.

It is one of Mr. Donnelly's contentions, in proof of the existence of Atlantis, that the horse, which, upon the evolutionist theory, he declares must have been first domesticated in America, could not have passed from America to Europe without the existence of " continuous land communication between the two continents."

Now, let us approach the question from the opposite direction. According to the biblical indications, and the tradition that mankind overspread the earth from the plains of Mesopotamia, we should expect to trace the possession and use of the horse. in the countries intermediate between the Tigris and Atlantic from East to West.

If, however, Atlantis existed, and was the original seat of civilisation and the point from which it spread to other countries, and if it is part of the statement that the horse existed on the island, then reversely we should expect to trace the progress of its use from West to East.

· M. Lenormant, it need scarcely be added, without any advertence to this question, has shown in his *Premières Civilisations* (p. 300),

" That the horse not only does not appear in any monument of the old empire, but is equally absent from those of the period called the middle empire, which extends from the first Egyptian revival under the eleventh dynasty until the invasion of the shepherds. . . . On the contrary, when the monuments recommence after a somewhat lengthened interruption under the eighteenth dynasty, the horse is seen as an animal in habitual use in Egypt."

On the other hand, the philological argument, the only one to which we can have recourse, would seem to show that the horse was well known in the East during the period it was absent from Egypt:

" The horse was one of the domestic species which the Aryans possessed in the earliest times, and the use of which was general among their tribes before they were dispersed, some in Europe, the others in Persia and India " (p. 318).

The evidence which we possess as to the migration of the horse appears to me decisive.

There is one other statement which I should like to have discussed, the only remaining one which has a look of corroboration of Mr. Donnelly's theory—*i.e.* apart from his diluvian traditions, which would drift us too far in their current. This chapter, however, has already run to too great length, and the statement will perhaps be more appropriately reserved for consideration in a subsequent chapter, in which I think I shall be able to disclose the secret of Atlantis.

CHAPTER II.

CONJECTURE AS TO THE PROBABLE BASIS OF PLATO'S ATLANTIS.

IN my last chapter I reserved an argument of Mr. Donnelly's for further consideration, and as it is based on one of the facts upon which he apparently obtains foothold—one of the islets or peaks, so to speak, of the submerged Atlantis —I will give it in extract:

"There was an ancient tradition among the Persians that the Phœnicians migrated from the shores of the Erythean Sea, and this has been supposed to mean the Persian Gulf; but there was a very old city of Erythia in utter ruin at the time of Strabo, which was built in some ancient age long before the founding of Gades, near the site of that town on the Atlantic coast of Spain. May not this town of Erythia have given its name to the adjacent sea? and this may have been the starting-point of the Phœnicians in their European migrations. It would even appear that there was an island of Erythia" (Donnelly's *Atlantis*, p. 310).

It will be perceived that this conjecture rests entirely on the statement of Strabo. In the first place, between Strabo's time and the commencement of Phœnician enterprise (B.C. 1200, Lenormant) there was full lapse of time for a city to have been founded, matured, and, the monarchical stage having elapsed, to have passed through the inevitable stages of aristocracy, democracy, despotism, revolution, and decay, and so in Strabo's time to have been entitled to the description of an ancient city.

But Strabo (the sole authority cited) himself says, according to Lenormant, without reference to this question:

" We must especially bear in mind the information preserved by Strabo (xvi. 766) with reference to the country first occupied by the Canaanites in *the Persian Gulf*, information which substantially agrees with that which Herodotus (I. i. v. 89 ; cf. Justin. xviii. 3) had collected from the mouths of the Phœnicians themselves, that the two most ancient sanctuaries of their race were situated in the islands of Tylos and Aradus (two of the existing Bahrien islands), which reproduced later on in the new country of the Phœnicians in the Mediterranean the island of Tyre and Aradus " (*Fragmens Cosmogoniques*, p. 221).*

Even if Strabo had not said it, another line of tradition would show that the Phœnicians sprang from the Erythean Sea, between the Red Sea and the Persian Gulf; but as this will afford evidence in another direction also, it will be convenient to reserve it. The evidence which has now accumulated will justify our reverting to Plato's fragment with a view to discover, if possible, what its real import may be.

Plato's *Atlantis*, so far as I know, has never been compared and confronted with a document, the authenticity of which is recognised by Heeren and Lenormant (it will be found *in extenso* in F. Lenormant, *Mém. d'Hist. Ancienne*, ii. 414, and also in Heeren, *Hist. Researches, Afric. Nations*, p. 478), viz. " the voyage of Hanno, which he has posted up ἀνεθηκεν in the temple of *Kronos*." The voyage of Hanno took place *circa* B.C. 500, and Plato was born *circa* B.C. 430. This document, which has come down to us in the form of a Greek translation, may reasonably

* M. Lenormant says this still more explicitly and emphatically (*Hist. Anc.* ii. p. 241), as if in anticipation of some such theory as that of Mr. Donnelly's. He says : " The Phœnician tradition, gathered at Tyre itself by Herodotus, . . . accepted equally by the judicious Trogus Pompeius ; the tradition of South Arabia, which Strabo has reported ; in fine, that which was current in the first centuries of the Christian era, when the original Syro-Chaldaic MS. of the book '*L'Agriculture Nabutienne*' was written, all three agree in declaring that the Chanaanites had primitively dwelt near the Chusites, their brethren in origin, upon the *shores of the Erythean Sea or Persian Gulf*." Further evidence is adduced, but this will perhaps suffice.

be presumed to have been accessible to Plato during his residence either in Sicily or in Cyrene.

It is my contention (1) that this document forms, so to speak, the backbone of the *Atlantis*. I think that I shall be able to show that Plato does not state any fact respecting Atlantis which has not been taken from this document except (2)—for I think the exceptions are sufficiently important to justify a second assertion respecting it—unless what Plato drew from the well of general or family tradition. Over the whole there is the glamour of Plato's style and imagination. Reserving what is preliminary, the account of Atlantis commences thus :*

" The tale, which was of great length, began as follows : I have before remarked, in speaking of the allotments of the gods, that they distributed the whole earth into portions. . . . And Poseidon, receiving for his lot the island of Atlantis, begat children by a mortal woman, and *settled them* in a part of the island which I will proceed to describe. On the side towards the sea, and in the centre of the whole island, there was a plain which is said to have been the fairest of all plains, and very fertile. Near the plain again, also in the centre of the island, at a distance of about fifty stadia, there was a mountain not very high on any side. In this mountain there dwelt one of the earth-born primeval men of that country whose name was Evenor, and he had a wife named Lucippe, and they had an only daughter who was called Cleito." (*Critias :* Professor Jowett's *Dialogues of Plato,* ii. 603).

This allotment of the earth corresponds to the tradition of Pheroneus, "the father of mankind" (Clemens Alex. i. 380), to whom the distribution of mankind was attributed, "*idem nationes distribuit*" (Hyginus, 143), and whom Plato calls "the first."

Hanno sailed about 500 B.C. with sixty vessels and thirty thousand colonists.

Assuming that *Atlantis* was idealised from the narra-

* In Appendix A and Appendix B I give *in extenso* the description of Atlantis in Plato's *Critias* (Jowett's trans.), and the translation of the *Periplus* of Hanno from Heeren's *Hist. Researches.*

tive of Hanno, Atlantis would be coextensive with the
Carthaginian empire, including the Canary and Fortunate
Islands. *Poseidon*, son of Kronos, was the tutelary god
of the Carthaginians, as witness Hamilcar's elaborate
sacrifice to him in the war with Gelon (*Juventus Mundi*,
p. 249); and Lenormant terms him " the Libyan Poseidon "

The occupation of Atlantis by *Poseidon*, and " his
begetting children by a mortal woman," and " *settling*
them" in a part of the island, may be conjecturally supposed to be the Carthaginian colonisation of the islands
mentioned in Hanno's narrative and of the mainland
beyond the mountains of Atlas; and this seems exactly
confirmed when we read in Heeren (p. 40), " The colonists
which Hanno carried out consisted, as we are expressly
informed, of Liby-Phœnicians, and were not chosen from
among the citizens of Carthage, but taken from the
country inhabitants."

This corresponds sufficiently. It will be noticed that
Plato, after the passage about Poseidon (as above), gives a
description of a plain, and Hanno's account commences
thus : " When we had passed the Pillars of Hercules on
our voyage, and had sailed beyond them for two days, we
founded the first city, which we named ' Thymiaterium.'
Below it lay an *extensive plain*." The passage in Plato
about Poseidon refers to the foundation of his first city.
As regards the derivation of " Thymiaterium," it is difficult to get beyond what old Bochart wrote, " Θυμιατήριον,
id est Thuribulum quorsum ?" Thymiaterium, Lenormant
tells us, is the modern " Mamoura "—Mamora. Now,
the description of Mamora very well corresponds with
Plato's descriptions. " It is situated upon a hill, near the
mouth of the river Suboe, the waters of which, gradually
widening in their course, fall into the Atlantic at this
place and form a harbour for small vessels." " The fer-

tile pastures, the extensive waters and plantations, which
we passed on our way hither have already been remarked."
" We travelled among trees of various kinds, so agreeably
arranged that the place had more the appearance of a
park than of an uncultivated country. We crossed plains
which were rich with verdure, and we had a view of lakes
which extended many miles in length."* McCulloch
(*Geog. Dict.*) says, " Morocco (the ancient Mauritania) has
a large extent of comparatively level land. Some of the
plains and valleys are of great extent and extraordinary
fertility ;" " the soil is now, as in antiquity, proverbial for
its fertility;" "the grass often attaining a height unequalled
except in the prairies of America." " On the north-
western side of the Atlas range the climate is healthy and
genial." Θυμιατήριον is only the Greek rendering of the
Libyan-Phœnician name, and perhaps a fanciful rendering. Bochart's† conjecture is that it was so called because
" situated in a plain," which corresponds to the fact ;
and Plato describes the plain in which Poseidon (Neptune) " settled his children " " as the fairest of all plains,
and very fertile."

Plato then proceeds abruptly to inform us that " Poseidon next, as he was a god, found no difficulty in making
special arrangements for the centre island, bringing the
streams of water under the earth, which he caused to
ascend as springs, one of warm water and the other of
cold ; and making every variety of food to spring up
abundantly on the earth." Here Plato a little anticipated
Hanno's narrative—apparently for the purpose of introducing the earliest Athenian legend concerning Poseidon,
for he is made to perform at Atlantis the same feat with

* Lemprière's *Tour to Morocco* (Pinkerton), xv.
† Bochart's conjecture is founded on a Hebrew equivalent; but this
may hold, as the Phœnician is classed by the philologists as a Shemitic
tongue. Concerning the extension of the Shemitic race, *vide Origin of
the Nations of Western Europe*, by J. Pym Yeatman.

which he is credited at Athens. "In his reign (Cecrops) Poseidon called forth with *his trident a well* on the Acropolis" (Smith's *Mythological Dictionary*).*

Hanno goes on to say that after passing the plain they proceeded first to the west, where, "in a place thickly covered with trees," they "erected a temple to Neptune" (Poseidon), and then to the east, "where we found a lake lying not far from the sea," which would correspond to "the lakes which extended many miles in length" (*supra*, p. 26). If they came upon a country where sea and land, land and lakes, alternated, it might have suggested to Plato's imagination "the alternate zones of sea and land." Plato says, "And we are further told that Poseidon, when he broke up the ground, ... made alternate zones of sea and land, larger and smaller; encircling one another."

It is next stated in Plato that Poseidon proceeded "to beget five pairs of male children, dividing the island of Atlantis into ten portions." "The eldest, who was the king, he named Atlas, and from him the whole island and the ocean received the name of Atlantic." The name of Atlas is here imported and transferred to the island by Plato from the traditions of Atlas on the mainland.

Then follows a long account of the settlement of the five pairs of male children, which might be allowed to pass and form the foundation for the theory of Atlantis, if, in *corresponding sequence*, Hanno had not added, "having passed the lake about a day's sail, we founded cities. . ." *Five* cities are named, the number corresponding with the five pairs of children of Poseidon.

Plato then descants upon the wealth and possessions of Atlas ; but before his eloquence has expended itself, he abruptly and incongruously says, as if in recollection of some fact, "Moreover, there were a great number of

* Compare *supra*, p. 14.

elephants in the island, and there was provision for *animals of every kind,* both for those that live in *lakes* and *marshes* and rivers, and also for those which live in mountains and on plains. . . ." In curious juxtaposition with this I may place Hanno's statement just before his mention of the five cities : " We proceeded until we arrived at a *lake* lying not far from the sea, and filled with abundance of *large reeds.* Here *elephants* and *a great number of other wild animals* were feeding."

The coincidence of the mention in both narratives, equally abruptly and unexpectedly, and in almost identical words, of elephants and other animals is noticeable, but there is another coincidence equally remarkable. Plato (p. 406) says : "The island in which the palace (the palace of Poseidon) was situated had a diameter of *five stadia.*" The Atlantis island, or continent, thus shrinks to these dimensions. No doubt there is mention of a *central* island, which implies others ; but the above gives us a measure of the localities indicated, which correspond very closely with the islands mentioned in Hanno's exploration.

Hanno says : " Thence we proceeded towards the east, the course of a day. Here we found in the recess of a certain bay a small island, containing a circle of *five stadia.*" " There we settled a colony, and called it Cerne." But this small island would appear to have been their head-quarters, for it is added, " We then came to a lake : . . . this lake had *three* islands larger than Cerne, whence, returning back, we came again to Cerne."

If Hanno's narrative lies at the foundation of Plato's fragment of Atlantis, it is natural that what is central in the one should be central in the others, and, accordingly, that what was the head-quarters in the one should figure as the palace of Poseidon in the other.

There is a slight resemblance in the way in which the

two narratives proceed. " Enough of the royal palace. Crossing the *water* harbours, which were *three* in number " (Plato). Hanno, after the mention of Cerne, which corresponds to the palace : " We then came to a lake, which we reached by sailing up a large river. This lake had *three* islands."

Several pages follow in Plato in description of the city—" the nature and arrangement of the rest of the country," and " the relations of their governments one to another "—to which nothing in the short narrative of Hanno corresponds, and for which the explanation must be sought elsewhere. (*Vide infra*, ch. v. p. 77.)

At the conclusion, however, of the two narratives there are descriptions which are very similar, and leave the impression of one having been suggested to the imagination by the perusal of the other.

Hanno says : " Towards the last day we approached somé *large mountains* covered with trees, the *wood* of which was *sweet-scented* and *variegated*. Having sailed by these mountains for two days, we came to an immense opening of the sea, on each side of which, towards the continent, was *a plain*, from which we saw by night fire arising at intervals in all directions, more or less ;" and further on, " When we had landed we could discover nothing in the daytime except trees ; but in the night we saw *many fires* burning, and heard the sound of pipes, cymbals, drums, and confused shouts. We were then afraid, and our diviners ordered us to abandon the island."

Plato describes Atlantis thus : " The whole country was described as being very *lofty* and *precipitous* on the side of the sea, but the country immediately about and surrounding the city was a *level plain*. . . . The surrounding mountains," " for their number, size, and beauty," " exceeded all that are now to be seen anywhere, having in them " . . . " *woods* of various sorts *abundant* for every *kind of work.*" " Also whatever *fragrant* things there are in the earth, whether roots or herbage or *woods*, grew and thrived in that land." After an account of their laws and customs, he describes their sacrifices of bulls to Poseidon —how they burnt the limbs of the bull, and took the rest of the victim to the *fire*, after having made a purification of the column all round, and then poured a libation on *the fire ;* and when darkness came on, and the fire about the sacrifice was cool (but not extinct), " all of them put on most beautiful azure robes, and, sitting on the ground at night near the *embers* of the sacrifice, on which they had sworn, and extinguishing all the fires about the temple, they received and gave judgment . . . ."—a scene which, if accompanied, as we may imagine, with " sound of pipes, cymbals, confused shouts," &c., would bring to the mind much the same scene which affrighted the mariners and diviners of Hanno's fleet.

Hanno's short narrative, or, at any rate, the Greek translation of it which has come down to us, omitting some final words about a savage people " whose bodies were hairy "—conjectured by Lenormant and others to be gorillas, that word having been wrongly substituted for the " gorgones or gorgades of the original MS."—may be said to end with a description of a volcanic region :

" Sailing quickly away thence, we passed a country burning with fires and perfumes ; and streams of fire supplied from it fell into the sea. The country was impassable on account of the heat. We sailed quickly thence, being much terrified ; and passing on for four days, we discovered at night a country full of fire. In the middle was a lofty fire, larger than the rest, which seemed to touch the stars. When day came, we discovered it to be a large hill, called the *chariot of the gods.*"

Plato's fragment—and it is a circumstance to be noted that both are fragmentary—terminates with the following passage, which, apart from the argument, may be acceptable :

" For many generations, as long as the divine nature lasted in them, they were obedient to the laws, and well affectioned towards the gods who were their kinsmen ; for they possessed true and in every way great spirits, practising gentleness and wisdom in the various chances of life, and in their intercourse with one another. . . . But when this divine portion began to fade away in them, then they, being unable to bear their fortune, became unseemly, and to him who had an eye to see they had lost the fairest of their precious gifts ; but to those who had no eye to see the true happiness they still appeared glorious and blessed at the very time when they were filled with unrighteous avarice and power. Zeus, the god of gods, who rules with law, and is able to see into such things, perceiving that an honourable race was in a most wretched state, and wanting to inflict punishment on them, that they might be chastened and improve, collected all the gods into his most holy habitation, which, being placed in the centre of the world, sees all that partake of generation. And when he had called them together, he spake as follows :"

There is nothing more, perhaps for the reason suggested ; for Hanno's narrative or the Greek translation extends no farther.

The catastrophe which was left thus vaguely impending had to be interpreted in the light of the previous statement (p. 599) that "Atlantis was sunk by an earthquake." Thus one narrative ends somewhat abruptly with the description of a volcano, and the other with a prognostication of a volcanic subsidence. If it were worth while, I might show a further coincidence in the approximation of the term used by Hanno, " the chariot of the gods," with the expression of Plato, " collecting all the gods into his most holy habitation."*

As I have said, there is nothing more ; but if I have succeeded in demonstrating that what is known as the

* Comp. also *supra*, p. 12, as to the centre of the world ; remark the striking resemblance to this description in the Chaldean account of the Deluge discovered by Mr. George Smith : (col. iii. 5-7) "The gods passed the tempest and sought refuge ; they ascended to the heaven of Assu ;" (17, 18) " The gods, in seats seated in lamentation, covered their lips for the coming evil."

Periplus of Hanno is the foundation of Plato's *Atlantis*,
the discovery, if I may so term it, will at any rate supply
the reason why the *Critias* (*Atlantis*) was never completed,
which has remained a difficulty even to Professor Jowett.

" The *Critias* [*Atlantis*] is a fragment which breaks off in the
middle of a sentence. . . . Why the *Critias* [*Atlantis*] was never
completed, whether from accident or advancing age, or from a
sense of the artistic difficulty of the design, cannot be determined "
(Professor Jowett's Introduction to Plato's *Dialogues*, ii. 595).

In speaking of the *Atlantis* as a fiction I by no means
intend that it was a fabrication intended to deceive his
contemporaries. It rather seems to me as if Plato was
indulging with them in a common and customary gratifi-
cation of the imagination, and that this is almost acknowledged in the following preliminary conversation :
" Consider, then, Socrates, if this narrative is suited to
the purpose, or whether we should seek for some other
instead." Socrates : " And what other, Critias, can we
find that will be better than this, which is *natural and
suitable to the festival* of the goddess, and has the advan-
tage of being a fact, and not a fiction?" (True in so far as
it was founded on Hanno.) " *How* or *where* shall we find
others if *we abandon* this ? There are none to be had "
(*Timæus*, 27 : Jowett). In other words, " I have brought
an interesting document from foreign parts, and if you
approve I will interweave it with our traditions."

CHAPTER III.

FURTHER CONJECTURES—DILUVIAN TRADITIONS.

In the last chapter I ventured to contend that the Periplus of Hanno was the main foundation for Plato's myth of Atlantis. Even, however, if this is conceded, something more will be required to dispel this "mirage" which has so long hung in the retrospect of human events.

Just as the "mirage" has led many in the past to their doom in the desert and in the ocean, so is it now apparently alluring them to abysses in the region of speculation.

"The fiction," says Professor Jowett, "has exercised great influence over the imagination of later ages. . . . Without regard to the description of Plato, and without a suspicion that the whole narrative is a fabrication, interpreters have looked for the spot in every part of the globe, America, Palestine, Arabia Felix, Ceylon, Sardinia, Sweden. The story has had also an effect on the early navigators of the sixteenth century " (ii. p. 590).

If Plato had spoken with full and exact knowledge of what was known in his days as to the extent of the exploration beyond the Pillars of Hercules, and had deliberately asserted his opinion as to the existence of "a lost continent," his opinion would have had great weight. But all that he says is that in consequence "of the subsidence of the Island" "the sea in those parts is impassable and impenetrable, because there is such a quantity of shallow mud in the way."

The manner in which he thus alludes to the Mare di Sargasso looks as if he had heard something of the explora-

tion, but his saying this and no more would also convey
the impression that he had heard of it traditionally, at any
rate not very directly. I notice that Sir J. Lubbock
(*Prehistoric Times*, p. 39) suggests that the existence of
this sea of seaweed itself originated the idea of the sunken
island. He says, "May not the belief in the 'Atlantis'
be as probably owing to the 'gulf-weed,' which would so
naturally suggest the idea of sunken land, as to any of the
other causes which are usually assigned for it?" And if
this "gulf-weed" formed an impassable and impenetrable
barrier to exploration in the Atlantic, it must have been
a constant subject of speculation with the Phœnician
mariners.

Although the conception of Atlantis arose as a myth in
the mind of Plato, there is every indication that a great
deal of floating tradition was used in its fabrication, and
this "residuum" will remain after the dispersal of the
"mirage."

There is one statement which strangely falls in with
the lines of tradition, and which can scarcely escape observation when attention is directed to Plato's narrative, viz.
that when he was ten years old, at a particular feast—
the Apaturia or "registration of youth"—he was told the
history of the Deluge. For whether it was a true or false
story of the Deluge, whether it was the universal Deluge,
or only the deluge which destroyed the island or "continent" of Atlantis, the fact remains as regards this discussion that it was either the Deluge which Moses and
the Hebrews and the Chaldeans and general tradition
record, or it was the subsidence of Atlantis which, according to Mr. Donnelly, lies at the foundation of all these
traditions.

Before proceeding in the inquiry it may be well to have
Plato's words before us :

"I will tell you an old-world story which I heard from an aged man; for Critias was, as he said, at that time nearly ninety years of age, and I was about ten years old. Now the day was *that day* of the Apaturia which is called the registration of youth. . . " He had previously referred to it as an ancient tradition—"he told us an ancient tradition" (Jowett's *Dialogues of Plato: Timæus*, p. 517).

These words taken in connection with the general tradition are very remarkable, but their full significance will not be appreciated until it is seen how closely the tradition in Ancient Greece resembles the diluvian traditions in America and Africa. It is not, however, my intention to recapitulate here the evidence which I have collected in chap. xi. of *Tradition*, and which, so far as I know, has not been rebutted, but to supplement it.

As, however, it may be rash to assume that the reader has read, or retains in his recollection, the curious ceremony commemorative of the Deluge which Catlin witnessed among the Mandan Indians in 1832, it will be necessary to give a few details. Mr. Catlin's account is attested by J. Kipp (agent to the Missouri Fur Company), J. Crawford Clark, and Abraham Bogard, who accompanied him; and in a subsequent account, published in 1867 by Messrs. Trübner, a letter of the Prince of Neuwied is printed, fully corroborating Mr. Catlin's statements from what he heard during a winter's residence among the Mandans, although he did not actually witness the ceremony; and in *Tradition*, p. 272, I pointed out that the ceremony among the Mandan Indians had been mentioned and briefly described in *Cérémonies Religieuses* a century before Catlin's visit to them.

That the Prince of Neuwied did not witness it is accounted for by the circumstance that it is only performed once a year, and in the spring.

" I resolved to await its approach," says Mr. Catlin, "and on inquiry found 'it would commence as soon as the *willow* leaves were full grown under the bank of the river.' I asked him why the

willow had anything to do with it, when he again replied, ' The twig
which the bird brought into the Big canoe was a willow hough,
and had full-grown leaves on it.' It will here be for the reader
to appreciate the surprise with which I met such a remark from
the lips of a wild man eighteen hundred miles from the nearest
civilisation." The ceremony in question, the O-kee-pa, Mr. Catlin
says, " though in many respects apparently so unlike it, was strictly
a *religious ceremony* [the italics are Mr. Catlin's], with abstinence,
with sacrifices, and with prayer, whilst there were three other distinct
and ostensible objects for which it was held. 1st. As an annual
celebration of the event of the ' subsiding of the waters ' of the
Deluge, of which they had a distinct tradition, and which in their
language they called *Mee-ne-ró-ka-hã-sha* (the settling down of the
waters). 2nd. For the purpose of dancing what they called the
Bull dance, to the strict performance of which they attributed the
coming of *buffaloes* to supply them with food during the ensuing
year. 3rd. For the purpose of conducting the young men who had
arrived at the age of manhood during the past year through an
ordeal of privation and bodily torture, which, while it was supposed
to harden their muscles and prepare them for extreme endurance,
enabled their chiefs, who were spectators of the scene, to decide
upon their comparative bodily strength and ability to endure priva-
tions and sufferings that often fall to the lot of Indian warriors, and
that they might decide who amongst the young men was the best
able to lead a war-party to an extreme exigency."—*O-kee-pa of the
Mandans*, by G. Catlin (Trübner & Co., 1867), p. 9.

Two facts, then, are in evidence : (1) that when
Plato was ten years old, at a feast called the "Apaturia"
or the "registration of youth," he heard a discourse
delivered which collected various diluvian traditions ; and
(2) that under strangely different circumstances of time
and place Catlin came upon a curious ceremony professedly
commemorative of a universal deluge, in which again a
principal feature or interlude was a ceremony which might
be exactly described as a registration of youth.

If, moreover, several other points of resemblance can
be shown between the Greek and Mandan festivals, this
discovery will go far to preclude any theory which would
account for the American tradition through local conditions
and modes of thought, and will further justify us in inter-

preting the one by the other, and regarding them as divergent lines of primitive tradition.*

In the latter ages of Greece the festivals were innumerable, more especially, as Xenophon tells us, among the Athenians ; but all that trace back to the remote past will be found upon analysis to be reducible to one or two primitive traditions.

Aristotle says, with a certain tone of authority which conveys the impression that he had in some way been behind the scenes, and knew the facts, that "the ancient sacrifices and festivals appear to have taken place after the ingathering of the crops, as first-fruits." Αἱ γὰρ ἀρχαῖαι Θυσίαι καὶ σύνοδοι φαίνονται γίνεσθαι μετὰ τὰς τῶν καρπῶν συγκομιδὰς οἷον ἀπαρχαί.—*Aristotelis Ethica Nicomachea*, viii. 11 (9). This view seems also to find expression in Virgil, *Eclogue* v. 90, where the rustics are made to invoke the primitive deities, Bacchus and Ceres (comp. also *infra*, p. 40), to whom "vota quot annis Agricolæ facient."

The later festivals were, as I have said, numerous ; but setting aside such as had only a local or historical origin, and confining the analysis to those which were professedly the most ancient, we come upon many features which confirm the statement of Aristotle—which accords with what we should have conjectured to be likely upon the scriptural indications in Genesis.† The more ancient festivals were

* Since I have written this chapter I have come upon the following passage in M. A. Reville's *Les Religions des Peuples non-civilisés*, i. 263 : " We find among the Redskins an institution which is very similar to the one we have seen in force in Africa, more especially among the Caffre-Hottentot groups, namely, a sort of religious and moral initiation of youth at the age at which the young man claims admission into the rank of warriors. These formalities are often very severe. Among the Dacotas, the Mandans," &c.

† There is a still greater correspondence with the Hebrew festival of the ingathering of the harvest on the fifteenth day of the seventh month : "And you shall take to you on the first day the fruits of the fairest trees . . . and the *willows* of the brook " (Leviticus xxiii. 39, 40).

held in honour of Jupiter, Apollo, Minerva, Bacchus (Dionysus), Neptune (Poseidon), Ceres, and Diana. I come to this conclusion upon the examination of the (*circa*) 319 festivals, the record of which Bishop Potter collects in his chapter on "Grecian Festivals," in his *Archæologia Græca*. He is also of opinion, basing it apparently on the passage already quoted, "that originally, as Aristotle reports, there were few or no festivals among the Ancients except those after harvest or vintage." If Apollo represented the Messianic tradition, we may see a special reason for the observance of his festival in the earliest times, and yet without anything specially commemorative of the Deluge. As I have discussed this question in the *Month*, April 1877, I omit further reference to it here; but in the other festivals above mentioned as primitive there will be found something which connects them with the diluvian tradition.

Jupiter ("Dyaus pater = Zeus pater = Jupiter," *vide* Max Müller and *Tradition*, p. 169), again, like Apollo, might have been expected to have had a festival apart; yet at any rate the culture secondarily became associated with the tradition, for in the curious annual festival the Hydrophoria, to which we shall again have to refer, the Athenians with great pomp carried vessels of water, which they poured into a gulf or opening in the temple of Jupiter; ". et dans cette occasion ils se rappaloient le triste souvenir que leur ancêtres avoient été submergés " (Boulanger, *L'Antiq. dévoilé par ses Usages*, i. 38). He adds they threw into the same chasm cakes of meal and honey (Pausanias, i. 18).

. This may be compared with the following incident in the *O-kee-pa* of the Mandans. The mysterious individual who opens the ceremony calls at each wigwam, and, " relating the destruction of all the human family by the Flood, excepting himself, who had been saved in his big

canoe, and now dwelt in the west," demands "some
edged tool to be given to the water as a sacrifice." "On
the last day of the ceremony, at sundown, in the presence of the chiefs and all the tribes," the tools were
thrown "into deep water from the top of the rocks, and
thus made a sacrifice to the water."

Zeus or Jupiter is more directly connected with the
Deluge, as the flood of Deucalion occurred because "Zeus
determined to destroy the human race by a great flood"
(Murray's *Myth.*, p. 42). I have not met with any refuta-
tion of the arguments identifying the deluge of Deucalion
with the universal Deluge (*ride Tradition*, p. 222 to p.
235). The same ceremony is described by Lucian at
Hierapolis in Syria, where there was the same custom
of pouring water into the cleft of the temple. Brett, in
The Indian Tribes of Guiana, gives a legend very similar
to that of Deucalion. Grote (*Hist. Greece*, i. 133) says,
"In this, as in other parts of Greece, the idea of the
Deukalionian deluge was blended with the religious im-
pressions of the people, and commemorated by their most
sacred ceremonies." Boulanger (i. 39) says, "It was,
according to the legend, by the opening of this chasm that
the waters which covered Attica had disappeared; and it
was alleged that Deucalion had erected an altar near this
place, and tradition attributed to Deucalion and his gratitude towards the gods the first foundation of the temple
of Jupiter Olympius, 'auprès duquel se faisoient ces cérémonies lugubres.'"

Assuming a simple primitive festival which formed the
"nucleus" round which the various diluvian traditions
collected, the prominence of the festivals of Ceres and
Diana would be respectively accounted for by the general
tradition having passed through a people who were either
husbandmen or hunters in their origin or in their predominant constituent. A pastoral people would have re-

mained more scattered and isolated, and their tradition
would require special consideration.

If, therefore, we find that iron was thrown into the
water as a token of sacrifice in one instance, and meal
in another, it would be only what we should expect in the
case of tribes having different avocations, but a common
tradition.

In Athens there was a feast called Ἀλῶα "in the
month of *Posidon,* in honour of Ceres *and* Bacchus, by
whose blessing the husbandmen received the recompense
of their toil and labour; and, therefore, their oblations
consisted of nothing but the fruits of the earth. Others
say this festival was instituted as a commemoration of the
primitive Greeks, who lived ἐν ταῖς ἀλῶσι, *i.e.* in vine-
yards and cornfields."* This festival recalls the primitive
simplicity of the ancient festivals noted by Aristotle, and
at the same time indicates a fusion with the diluvian
traditions in its connections with Posidon and Bacchus.
There was another festival (p. 400) named from "the
gathering of the fruits," held, according to Menander, in
honour of Ceres and Bacchus, and at which, according to
Eustathius, "there was also a solemn procession in honour
of *Neptune.*" We are elsewhere told that the festival
θεσμοφόρια was kept universally throughout Greece, except
by the Eretrians (p. 404). This festival was held in honour
of Ceres, as the law-giver, "because she was the first who
taught mankind the use of laws;" which may mean that
the festival went back to the time when law commenced
or recommenced—to Noah and the Deluge.†

There was one sacrifice in the Aloa festival which has

* Bishop Potter, i. p. 361.

† " On peut voir ici une application de la règle d'après laquelle on
trouve l'origine des coutumes les plus bizarres, quand on peut les com-
parer chez les peuples divers, entourées et comme flanquées de coutumes
nécessoires moins fréquemment observables, variant d'un peuple à l'autre,
mais pivotant autour d'une idée toujours la même " (*Les Religions des
Peuples non-civilisés,* par A. Réville, 1883, i. p. 337).

a close resemblance to a festival among the Minatarees
—the village community adjoining the Mandans—the
Thalusia—"a sacrifice offered by the husbandmen after
harvest, ὑπὲρ τῆς καρποφορίας, i.e. in gratitude to the gods,
by whose blessing·they enjoyed the fruits of the ground.
The whole festival was called Aloa. . . . Hence comes
Θαλύσιος ἄρτος, sometimes called Θάργηλος, which was
the first bread made of the new corn" (Bishop Potter, i.
p. 400).

Compare Catlin, North American Indians, i. p. 189:

"At the usual season and the time when, from outward appearance of the stalks and ears of the corn, it is supposed to be nearly
ready for use, several of the old women who are the owners of fields or
patches of corn . are delegated by 'the medicine-men' to look
at the cornfields every morning at sunrise, and bring into the
council-house several ears of corn, the husks of which the women are
not allowed to break open, or even to peep through.. . . . When from
repeated examination they come to the decision that it will do, they
despatch runners or criers announcing to every part of the village or
tribe that the Great Spirit has been kind to them, and they must
all meet the next day to return thanks for his goodness."

A feast and dance follow. I will note further that
just as there was a festival in honour of Ceres when the
new corn* was ripe, so was there (Bishop Potter, p. 416) in
honour of Bacchus when the new wine was first tasted;
and another (p. 427), the Protrugeia, in honour of Neptune
and Bacchus in connection with the new wine. It will
be remembered that the Mandan diluvian commemoration
took place as soon as the willow leaves were full grown;
and at Athens (p. 393) there was a festival of elenophoria,
"from ἔλεναι, vessels made of bulrushes, with ears of
willow, in which certain mysterious things were carried
upon this day." The festival of Mysia (Potter, p. 415),
"in honour of Ceres, continued seven days, upon the

* The Hebrews (Leviticns xxiii. 10) were commanded " to bring sheaves
of ears, the first-fruits of your harvest, to the priests."

third of which, all the men and dogs being shut out of the temple, the women .. .. . remained within. . . ." In the Mandan festival, "orders were given by the chiefs that the women and children should all be silent and retire within their wigwams, and their dogs all to be muzzled during the whole of that day, which belonged to the Great Spirit" (Catlin, p. 11).

"In the middle of the last dance on the fourth day [in the Mandan ceremony, Catlin, p. 22], a sudden alarm throughout the group announced the arrival of a strange character from the West. . . . This strange and frightful character, whom they called the evil spirit, darted through the crowd when the buffalo-dance was proceeding. His body was painted jet black. . . ." He is confronted by the conductor of ceremonies and his medicine pipe, who, "looking him full in the face, held him motionless under its charm until the women and children had withdrawn from his reach." After a while the women gradually advanced and gathered around him. "In this distressing dilemma he was approached by an old matron, who came up slyly behind him with both hands full of yellow dirt, which, by reaching round him, she suddenly dashed in his face, changing his colour; . . . and at length another snatched his wand from his hand and broke it across her knee. . . . His power was thus gone, ... and bolting through the crowd he made his way to the prairies." In this we seem to see trace of the primitive tradition that the woman should crush the head of the serpent. This tradition,* which may be almost said to find direct expression in the antagonism of the serpent Python to Latona, and his final discomfiture and death at the hands of her son Apollo immediately upon his birth, may perhaps also be seen in the prominence given to women in some of the Grecian festivals—ostensibly, no

* Vide supplemental evidence, infra, p. 70.

doubt, in commemoration of some local victory; *e.g.*
(Potter, 404) : " There was a mysterious sacrifice called
diorma, or apodiorma, because all men were excluded,
because in a dangerous war the women's prayers were so
prevalent with the gods that their enemies were defeated
and put to flight as far as Chalcis ;" and in the utristika
at Argos (p. 435), " where the chief ceremony was that
the *men and women exchanged habits*, in memory of the
generous achievement of Talasilla, who, having enlisted
a sufficient number of women, made a vigorous defence
against the whole Spartan army." It should have been
mentioned that after the defeat of the evil spirit by the
woman in the Mandan ceremony, " the whole government
of the Mandans was then in the hands of one woman—she
who had disarmed the evil spirit; . . . that all must
repair to their wigwams ; . . . that the chiefs on that
night were *old women*, and had nothing to say. . . . "In
the Διονύσια ἀρχαιότερα (Potter, p. 383), as distinguished from the νεώτερα celebrated in the temple of
Bacchus, "the chief persons who officiated were fourteen
women, appointed by the Βασιλεὸς, who was one of the
Archons. . . . They were called the Venerable. . . ."

The Apaturia, it will be remembered, was the feast of
the "registration of youth," at which Plato tells us he was
told the legend of the subsidence of Atlantis, which, as I
contend, was only a form of the tradition of the universal Deluge. Now, the term "Apaturia," which signifies
"deceit" (Smith, *Myth. Dict.*), has no explanation in
anything that occurred or is recorded of the Grecian
festival.

There are, however, two legends in explanation. In
the one it is connected with a surname of Aphrodite, who
enticed the giants into a cavern to their destruction by
Heracles. The legend, no doubt, is susceptible of another
interpretation ; but in its main feature it is the destruction

of the spirits of evil through the "artifice of a woman,"[*]
In the other it is told that the festival was first instituted
at Athens in memory of the stratagem by which Melanthius, the Athenian king, overcame Xanthus, the King of
Bœotia, in single combat. As they were just going to
begin the fight, " Melanthius, thinking or pretending that
he saw at Xanthus's back a person habited in a *black* goat-
skin, cried out that the articles were violated ; upon this,
Xanthus, looking back, was treacherously slain by Melanthius " (Bishop Potter, i. 369). This brings to recollection the scene we have just witnessed, in which the Mandan maiden discomfited the evil spirit painted black by
stealthily approaching him from behind.

The resemblance might be deemed insufficient and
inconclusive, if it were not for the relation of the legend to
the festival of the " registration of youth ;" for this juxtaposition will be found also in the Mandan ceremony.
When the heroine after her victory is conducted to the
" medicine (or mystery) lodge," she orders the bull-dance
to be stopped, the four tortoise drums (concerning which
presently) to be carried in, the buffalo and human skulls
to be hung on the four posts, and *she* then invites the
chiefs to enter the medicine-lodge " to witness the voluntary tortures of the young men now to commence " (see
above, p. 36).

The Apaturia, it is true, did not, at any rate in the
time of Plato, present the horrible features of the Mandan
ceremony ; but in other Grecian festivals there are evidences of scenes quite as revolting as those which Catlin
witnessed, and enacted apparently upon the same motives

---

[*] In the first instance it is the alarm of the woman in the Mandan
ceremony which brings about the intervention of the man with the medicine or mystery pipe, who curbs the evil one, as in the legend Aphrodite decoys the giants to the caverns in which Heracles (who, according
to an interpretation of certain legends as Hercules might be called " the
first or only man," like the mystery man) is concealed.

and ideas. One idea seems common to them all—to provide a certain registration of youth; at Athens perhaps only a civil registration engrafted on a primitive festival, and at Sparta and among the Mandans a registration and test of fortitude and endurance.

The Mandan (Catlin, p. 28 to p. 31).	*The Spartan* (*Archæo. Græca*, i. p. 379).
The Mandans were suspended by splints inserted in the flesh until life was apparently extinct. "No one was allowed to offer them aid whilst they lay in this condition. They were here enjoying their inestimable privilege of voluntarily entrusting their lives to the keeping of the Great Spirit. . . . The young men seemed to take no care or notice of the wounds thus made. . . . During the whole time of this cruel part of the ceremonies the chiefs and other dignitaries of the tribes were looking on to decide who amongst the young men were the hardiest, who could hang the longest by his torn flesh without fainting, . . . that they might decide whom to appoint to lead a war-party, or to place at the most important posts in time of war." If death ensued, "they all seemed to speak of this as an enviable fate rather than as a misfortune; for the Great Spirit had so willed it for some especial purpose, and no doubt for the young man's benefit."	At Sparta it took the form of the flagellation of youths before the altar of Diana Orthia, and lest the youths "should faint under correction, or do anything unworthy of Laconian education, their parents were usually present, to exhort them to bear whatever was inflicted upon them with patience and constancy. And so great was the bravery and resolution of the boys, that though they were lashed till the blood gushed out, and sometimes to death,* yet a cry or groan was seldom or never heard to proceed from any of them. Those of them that died by this means were buried with garlands on their heads in token of joy or victory, and had the honour of a public funeral. . . . By some it is said to have been one of Lycurgus' institutions to accustom the youth to endure pain. . . ." By some it is traced to the introduction of the worship of Diana *Taurica*, and in mitigation of the oracle which commanded that human blood should be shed upon her altar.

Note in connection with the worship of Diana *Taurica* that the Mandan custom was preceded by "the bull-dance" and followed by "the feast of the buffaloes," and that the grand operator in the tortures sat with "a dried buffalo skull before him." In Sicily (p. 431) there was a festival in which the youths beat each other with sea-onions; the victor was rewarded with a bulb. I have made a suggestion as to the significance of the bull in connection with the diluvian tradition, in *Nature Myth Theory,* pp. 7-10, which I now reprint in Appendix C. The Lacedæmonians detested the worship of Diana Taurica, but feared the anger of the goddess. To the faithful observance of their custom the Mandans looked for their annual supply of buffaloes. There was a festival of Pan in Arcadia, when the boys used to beat his statue with sea-onions, more especially when they missed their prey in hunting.

There is something in the Mandan ceremony that reminds us of the Dionysia, although they would appear to have been a water-drinking people when Catlin visited them; at any rate, there is no mention of intoxicating drinks. The immorality of the closing scene in the ceremony, however, recalls the Bacchanalian orgies, and moreover the central object in their village, which they called "the big canoe," and round which the dances took place, was shaped like a hogshead cask (compare *infra*, p. 72). Assuming the fact that Bacchus represents the later traditions of the patriarch Noah, or possibly the tradition of Cham, embodying traditions of the episode recorded in the Bible, the substitution of wine-sacks for water-sacks in the following narratives would correspond to the confusion of tradition we have just seen in the combination of the wine-butt and the canoe.*

---

* The late Colonel George Macdonell, C.B., related that certain Jesuit missionaries went in search of an Indian tribe whom Sir John

Mandan (Catlin).

"There were also four articles of veneration. and importance lying on the ground, which were sacks containing each some three or four gallons of water. They seemed to be objects of great superstitious regard. . . . The sacks of water had the appearance of great antiquity, and the Mandans pretended that the water had been contained in them ever since the Deluge. . . . During each and every one of these bull-dances the four old men who were beating on the sacks of water were chanting forth their supplications to the Great Spirit for a continuation of his favours in sending them buffaloes to supply them with food for the ensuing year. . . ."

Archæo. Græca, i. p. 372.

"'Ασκώλια. A festival celebrated by the Athenian husbandmen in honour of Bacchus, to whom they sacrificed a he-goat because that animal destroys the vines. . . . Out of the victim's skin it was customary to make a sack, which, being filled with wine and oil, they endeavoured to leap upon it with one foot, and he that first fixed himself upon it was declared victor, and received the sack as a reward. The festival was so called from leaping on the sack (or bottle)." This must be considered in connection with the *conjoint* festivals of *Neptune* and Bacchus, *e.g.* the προτρυγεῖα, from "new wine."

The young men at Rome were invested with the *toga virilis* at the *liberalia*, a festival in honour of Bacchus (Döllinger, *Jew and Gentile,* ii. 51).

Ross in his voyage towards the North Pole had described as without any creed of any kind. The Jesuits found that they had no worship except that at midday they assembled in a circle, and then the oldest man called out three times "Ye-ho-wah," which they regarded as an invocation of Jehovah. His informant was the Rev. G. Glover, S.J., at Rome.

I find a very similar account in Stanley Faber's *Pagan Idolatry,* ii. p. 309, who quotes from *The History of the American Indians,* by James Adair, a trader with the Indians, and resident in the country for forty years. Mr. Adair gives an account of an Indian tribe who had carried about with them an ark in which they kept various holy vessels. "This ark the priests were wont to bear in solemn processions. They never placed it on the ground; when stones were to hand, they rested it upon them; when not, upon logs of wood. . . . No one presumed to touch it except the chieftain and his attendants, and only on particular occasions." The deity of this ark they invoked by the name of Yo-he-wah, which Mr. Adair supposes to be a slight variation of the Jehovah of the Hebrews. Faber, however, after adducing supplementary evidence, con-

If Dionysus (or Bacchus) embodies a tradition of Noah, and if (Gen. ix. 3) the permission to eat flesh-meat was first given to the patriarch, this is an event which we should expect to find transmitted in tradition, and we seem to see it in the Λειώνια and the ὠμοφάγια, festivals held in honour of Bacchus as "the eater of raw flesh" (*Archæo. Grœca*, 362-439).

The probability of such indirect tradition is increased by the fact of direct tradition in the pages of Porphyry, the opponent of Christianity (*vide* extract from Porphyry, *De Abstinentia*, liv. ii., in L'Abbé Gainet's *Hist. de l'Anc. et Nou. Test. par les seuls témoignages profanes*, i. 175). The aim of Porphyry's work was to revive the system of Pythagoras, and beyond it to bring men back to the manners of primitive life. " Now," he says, " these men, *les habitants voisins des générations divines*, eat nothing which had life, in order to give themselves up more freely to the exercise of the intellect, and to hold themselves aloof from the depravation of manners." Porphyry quotes Dicearchus to this effect, and adds, what has a special significance with reference to the theory of primitive barbarism: "And it is evident that this light and simple kind of food gave birth to the proverb which circulated in the succeeding ages, ' Then the acorn sufficed. . . .' "

We have seen at the commencement of this inquiry

---

cludes : " I am inclined to believe that as Ho is Hu or Bacchus (comp. Welsh Celtic legend of Hu, Faber, p. 304, and Chaldaic Hoa), so we have no other than the Bacchic cry of Hevah or Evoë, and consequently that the exclamation Yo-he-wah is in fact nothing more than Ho-Hevah, which is equivalent to Huss Evoe, or inversely Evoë Bacche." I must leave the reader to decide between these conflicting views. St. Clemens of Alexandria says, describing the orgies of the Bacchantes : " Coronati serpentibus et ululantes *Evam*. Evam illam per quam error est consecutens ; et signum Bacchicorum orgionem est serpens mysteriis initiatus " (*Admon. ad Gentis*, p. 9). I have discussed the evidence further in chapter iv. ; and also, with reference to the Indian snake-dance, compare page 71, *infra*.

that there was at Athens an annual ceremony directly commemorative of the Deluge, which was called the Hydrophoria, from bearing water, which they poured into an aperture in the temple, " in memory of those who perished in the Deluge." This ceremony has this feature in common with the Syrian ceremony described by Lucian. It is necessary to recall these facts in order to perceive the full significance of the objects carried in the procession (*vide Tradition*, p. 248). It was celebrated on the 25th Thargelion=5th May (comp. Catlin above). Every citizen contributed an *ox* and *olive*-branches. " In the ceremonies without the city there was an engine built in the *form of a ship* on purpose for this solemnity;" upon this the sacred garment of Minerva " was hung in the manner of a sail;" " the whole conveyed to the temple of Ceres Eleusinia." " This procession was led by old men, together, as some say, with old *women*, carrying olive-branches in their hands." " After them . . . . sojourners, who carried *little boats* as a token of their being foreigners, and were called on that account *boat-bearers;* then followed the women, ... who were named ὑδριαφόροι, from *bearing water-pots*" (comp. the hydrophoria, as above).

I have excluded from the inquiry two things which, in certain aspects, are common to all these ceremonies in their later and degenerate forms—obscenity and solar-worship. Before proceeding to justify the exclusion of the latter, I must conclude the evidence with one striking resemblance or coincidence between the Mandan and Grecian festivals, which perhaps ought to have been mentioned before. The opening of the Mandan ceremony is thus graphically described by Catlin (p. 9) :

" The season having arrived for the holding of these ceremonies, the leading medicine (mystery) man presented himself on top of a wigwam one morning before sunrise, and haranguing the people, told them that he ' discovered something very strange in the

E

western horizon, and he believed that at the rising of the sun a great white man would enter the village from the west and open the medicine lodge.' In a few moments the tops of the wigwams and all other elevations were covered with men, women, and children on the look-out; and at the moment the rays of the sun shed their first light . . . all eyes were directed to the prairie, where, at the distance of a mile or so from the village, a solitary human figure was seen descending the prairie hills and approaching the village in a straight line until he reached the picket. . . . The head chief and the council of chiefs . . . soon made their appearance in a body at the picket, and recognised the visitor as an old acquaintance whom they addressed as ' Nuh-mohk-múck-a-nah ' (the first or only man). All shook hands with him, and invited him within the picket. He then harangued them for a few minutes, reminding them that every human being on the surface of the earth had been destroyed by the water except himself, who had landed on a high mountain in the west in his canoe, where he still resides, and from whence he had come to open the medicine lodge, that the Mandans might celebrate the *subsiding of the waters* (comp. hydrophoria—Lucian's account of the ceremony at Hierapolis) and make the proper *sacrifices to the water*, lest the same calamity should again happen to them."

Let us listen if we do not seem to catch the echo of this in the account which Athenæus (lib. iii.) has given us of the πελώρια : "Baton, the Sinopensian rhetorician, in his description of Thessaly and Hæmonia, declares that the *Saturnalia* was a Grecian festival, and called by the Thessalians Peloria." His words are these :

" At a time when the Pelasgians were offering public sacrifices, one Pelorus *came in*, and *told one* of them that the mountains of Tempe in Hæmonia were torn asunder by an earthquake, and the lake which had previously covered the adjacent valley, making its way through the breach and falling into the stream of Peneus, had left behind a vast but most pleasant and delightful plain. The Pelasgians hugged Pelorus for his news, and *invited him to an entertainment* where he was treated with all sorts of dainties. . . . In memory of this, when the Pelasgians had seated themselves in the newly discovered country, they instituted a festival wherein they offered sacrifices to *Jupiter*, surnamed Pelor, and made sumptuous entertainments, whereto they invited not only all the foreigners among them (compare the Panathenæa), but prisoners also . . . and slaves, all of whom they permitted to sit down, and waited on them."

This latter feature connects it with other Grecian festivals, and also with the Persian festival held in the spring of every year, " when the husbandmen were admitted without distinction to the table of the king and his satraps " (Gibbon, ii. 8).

Considering the prominent part which Bacchus plays in these ceremonies and in mythological tradition, it is curious that he should only once be mentioned in Sir G. W. Cox's *Mythology of the Aryan Nations*. Is it not a notable instance of the part of Hamlet being omitted in the play of *Hamlet ?* This casual reference to him will be found (ii. p. 4), " Bacchos the son of Dionysos ;" and I may here notice that Sir G. W. Cox's mythology does not accord either with the discarded Lempriere, or with Dr. Smith's *Dictionary of Mythology*, or with Murray's *Manual*, even when they all agree, and Sir G. W. Cox gives few classical references. However, it is not necessary here to discuss whether or not " Bacchos " is " the son of Dionysos," or whether he is Bacchus the Roman equivalent for the Greek Dionysos.

I am not impeaching Sir G. W. Cox's extensive acquaintance with classical literature, but I do protest against the manner in which this school presents its theory to the exclusion of every other view. I think we might have expected some recognition of the theory originally held that Bacchus or Dionysos embodied traditions of the patriarch Noah or Cham or Nimrod (for either might have formed a " nucleus " round which the traditions or myths might have collected); at any rate, we should have expected some advertence to the facts.

The salient facts from the traditional point of view are:

1. That although there are legends of a youthful

Bacchus,* and of his double birth (equally explicable on
solar theory and on the theory of the revival of life and
second birth after the Deluge, which legend, by the bye,
takes the form of Adonis, who is also saved in an ark), yet
the circumstances and surroundings which the legends
reveal necessarily locate him in the primitive ages of the
world.

2. Neither is it a difficulty that there "were three
Bacchuses," for they are all resolvable into various forms
of the same legend, or myth, equally from the historical
or nature myth point of view. "Like the Theban wine-
god, Adonis is born only on the death of his mother; and
the two myths are, in one version, so far the same that
Dionysos, like Adonis, is placed in a chest which, being
cast into the sea, is carried to Brasiai, where the body of
his mother is buried" (Cox, *Myth.* ii. 9). "Adonis
stands to Dionysos in the relation of Helios to Phoibos"
(Cox, ii. 113). The extinct mother may as well be the
former world destroyed as the night or the winter.

3. Various legends connect Bacchus with the sea.

4. He is the first legislator.

5. Bacchus first discovered and planted the vine.

The principal traditions regarding Bacchus are recorded
by Diodorus Siculus, and although Diodorus explicitly
states this, the statement is ignored in Cox, Murray, and
even in Smith and Lempriere. This is so important that
I must give the actual text. Writing at a period B.C. 8,
when tradition had become obscured, Diodorus inclines to
the naturalist views, which, by the bye, presuppose the ante-
cedence of the historical tradition or myth, inasmuch as

---

* Diodorus Sic. tells us (l. iv.) that Dionysus (like Janus) had two
faces, "that the ancient Dionysus always wore a long beard." In the
other aspect he is represented "as a spruce young man." Janus (bifrons)
is represented "with a prow of a ship on the reverse of his medals,"
and on the Sicilian coins at Eryx with "a dove encircled with a crown,
which seems to be of olive" (Bryant, *Myth.* ii. 254).

they are critical attempts to explain them away, and do not materially differ from the modern attempts; *e.g.* Diodorus (i. 3) says, "And these are the opinions of those who take Bacchus for nothing else than the use and strength found to lie in the vine;" and in Cox (ii. p. 293), Dionysos (Bacchus) is apparently "the manifestation of that power which ripens the fruits of the earth, and more especially the vine." Having, however, given the naturalist view, Diodorus then says, "But 'those writers on mythology' who say that this god was a man *unanimously* attribute to him the *finding out* and *first planting* of the vine, and everything that belongs to the use of the vine" (Booth, tran. iii. p. 204). τῶν δὲ μυθογράφων οἳ σωματοειδῆ τὸν θεὸν παρεισάγοντες τὴν μὲν εὕρεσιν τῆς ἀμπέλου καὶ φυτείαν καὶ πᾶσαν τὴν περὶ τὸν οἶνον πραγματείαν συμφώνως αὐτῷ προσάπτουσι (Diodorus Siculus, lib. iii. c. 63). I shall return to this text in a moment.

6. Horace uses in respect to Bacchus the very traditional phrase *Father* Bacchus: "Quis non te potius Bacche pater" (*Odes*, i. 18).

7. Bacchus is described "as relentless in punishing all want of respect for his divinity," and the punishment of Pentheus, as narrated by Theocritus, B.C. 282, forcibly reminds us of the curse of Chanaan in Genesis:

"Perched on the sheer cliff, Pentheus would espy
　All . . ."

(For profaning thus "these mysteries weird, that must not be profaned by vulgar eyes," Pentheus is torn to pieces by the Bacchanals):

. . . "Warned by this tale, let no man dare defy
　Great Bacchus, lest a death more awful *should* he die,
And, when he counts *nine* years or scarcely *ten*,
　Rush to his ruin. May I pass my days
Uprightly, and be loved by upright men!
　And take this motto, all who covet praise

('Twas ægis-bearing Jove that spoke it first) :
The *godly seed* fares well, the wicked is *accurst.*"
                    (Calverley's *Theocritus*, Idyll xxvi.)

8. Bacchus was regarded as the god of the drama and
the protector of theatres ; very naturally, if the commemorative ceremonies we have been discussing were primitive
and anteceded the stage and the drama.

9. In the lines of the Orphic hymn :

Κικλήσκω Διόνυσον ἐρίβρομον εὐαστῆρα
πρωτόγονον, Διφυῆ τρίγονον.

The three last epithets exactly apply to Noah. Διφυῆ ==
double == bifrons, as we have just seen him represented
looking back on the world that had perished, and forward to the new birth he was inaugurating; τρίγονον in
allusion to his three sons; but more especially in the
epithet first-born — πρωτόγονον. As I have elsewhere
remarked, since all antediluvian traditions meet in Noah,
and are transmitted through him, there is an *a priori*
probability that we shall find all the antediluvian traditions confused in the tradition of Noah, and accordingly
the reduplication of Adam in Noah has not escaped observation. Thus the epithet "first-born" curiously applies; and
I must further note the remarkable resemblance to the
epithet addressed to the personage who appeared from the
plain to open the Mandan ceremony (*vide sup.*, p. 49),
" the first or only man."

The cumulative force of the evidence appears to me
very strong, and assuming the existence of Noah, I think
it must be admitted that the tradition points to the
patriarch, or to some original progenitor immediately in
contact with him. It may, indeed, be alternatively asserted that Bacchus is the sun, but this will not suffice.
This may be objected by those who believe in Genesis, or
by those who do not. If by the former, I reply that
whenever mythology commenced, Noah or Cham . may

have been identified with, or deified in connection with, the sun ; and from this point of view the mythology must have been subsequent to the Deluge, and, accepting the scriptural indications, nothing is more probable than that the mythology should have absorbed or embodied the incidents of that stupendóus event, and the personages and facts of early history.

To those who only recognise the book of Genesis as a historical record I must submit that in supposing the patriarch—perhaps they will recognise him, at any rate, as a progenitor—to have been deified in connection with the sun, I am not hazarding a mere conjecture, but am stating as a hypothesis what there is evidence to show was probable, because in accordance with the tendency of thought at that day.

"The Egyptian priests, as we learn from Plutarch (*De Isid.* p. 354, τὰς δὲ ψυχὰς λάμπειν ἄστρα), taught expressly that Cronos, Osiris, Horus, and all their principal deities were once mere men ; but that after they died their souls migrated into some one or other of the heavenly bodies, and became the genii or animating spirits of their new celestial mansions. . . . In a similar manner we are told by Sanchoniathon that Ilus or Cronos [comp. *supra*, p. 18] was once a man, that he was deified by the Phœnicians after his death, and that his soul was believed to have passed into the planet which bears his name. Eusebius, *Præp. Evan.* lib. i. c. 10 " (Stanley Faber, *Pagan Idolatry*, ii. p. 227).

The expression of Cicero also, " oportet contra illos etiam qui hos deos ex hominum genere in cœlum " (*De Nat. Deorum*, iii. 21), may be adduced in evidence of the tendency to this mode of deification. The inquiry would seem to have resulted, so far, in the following facts : Taking the evidence of Genesis (whether as revealed truth or historical record), at the primitive commencement of the human family after the Deluge (or of a large section of mankind after a deluge), we have the account of the first planting of the vine by the first progenitor, followed

by a scene of intoxication which would, from its circumstances, have impressed itself on the memories of his
descendants.

Tracing backwards, we find evidences of commemorative ceremonies, which correspond to the facts above recorded in the circumstances that they were commemorative
of a deluge, of a first legislator, of the first planting of the
vine, of the first cultivator, almost invariably terminating
in a scene of riot or intoxication—not always, it is true,
combining all these traditions, but so combining features
of them as to disclose a common parentage.

On the other hand, it is contended that the Deluge of
Genesis was only one line of the tradition of the submersion of Atlantis, which would give a measure of its mag-
nitude—" that great deluge of all," as Plato calls it. As
against this, I have pointed to a document to which
Plato's narrative very closely corresponds, and which, if
his narrative is not to be accepted literally, might easily
have formed the foundation for the fiction. That is how
the argument at present stands.

CHAPTER IV.

SINCE the previous chapter was written I have become acquainted with Mr. A. Lang's *Custom and Myth*, quite recently published (Longmans, 1884). Mr. Lang, having made the discovery that, in what is called the "orthodox" school of mythology—the school of Mr. Max Müller and Sir G. W. Cox—"the distinguished scholars and mythologists" "usually differ from each other," and manifest "none of the beautiful unanimity of orthodoxy," has wisely sought to broaden the basis of mythology. Instead of endeavouring to find its derivation in etymology, after the manner of the philological school, he rather seeks it in folklore, which, I may observe in passing, was the aim, as is implied in the title of my book, *Tradition with Reference to Mythology*, printed in 1872. Mr. Lang says (p. 25):

"Our method throughout will be to place the· usage or myth, which is unintelligible when found among a civilised race, beside the similar myth, which is intelligible enough when it is found among savages. . . . The conclusion will usually be that the fact which puzzles us by its presence in civilisation is a relic surviving from the time when the ancestors of a civilised race were in a state of savagery."

It will be noticed that Mr. Lang's view necessarily supposes a state of primitive barbarism. Now if, on the other hand, we point to the ancient and contrary belief of a large section of mankind that the race commenced with civilisation and with survival from a Deluge ; and if we further contend that, if this belief is accepted, all these myths and customs fall into their place, and that their similarity of feature demands no other explanation than community of origin, all I need say is that this belief, or theory, or tradition, cannot be displaced or overthrown by any theory which assumes the state of primeval barbarism, as this merely begs the question at issue. I feel dispensed, therefore, from further advertence to Mr. Lang's theory, and may confine myself to his facts, which bring striking corroborative evidence.

Mr. Lang's account of " the bull-roarer " brings fresh evidence in relation to the Mandan ceremony we have just been examining, proving, if need be, that Catlin's narrative is attested by external testimony, and further connecting it with the diluvian tradition.

The " bull-roarer " in itself need alarm no one, and an interesting account of this traditional toy will be found in Mr. Lang's book at p. 30. Its significance is in its relation to the diluvian ceremonies. It is simply a piece of pointed wood tied to a string, which, when whirled round, produces a roaring noise. Assuming the identity of bull-roarers, *turbines*, κῶνοι and ῥόμβοι, the latter being sometimes interpreted as " a magic wheel," we may assume also their identity with the " rattles " used in the Mandan ceremony (*infra*, p. 67).

The " bull-roarer," like the " rattles," " is associated with mysteries and initiations." If the belated traveller in the plains of Australia hears the bull-roarer, ". he knows that the blacks are celebrating their tribal mysteries." " The roaring noise is made to warn all women to keep out of the way " (compare *sup.* p. 41) ; " just as Pentheus was killed (with the approval of Theocritus) because he profaned the rites of the women-worshippers of Dionysus " (compare *sup.* p. 53).

Mr. Lang adds (p. 34) :

"Among the Kurmai in Australia the sacred mystery of the 'turndun,' or 'bull-roarer,' is preserved by a legend which gives a supernatural sanction to secrecy. When boys go through the mystic ceremonies of *initiation*, they are shown the bull-roarers, and made to listen to their hideous din. They are then told that, if ever a woman is allowed to see a 'turndun,' the earth will open, and *water will cover the globe.*"

Here we have the "turndun" connected with the tradition of the Deluge, and probably also with the tradition of Eve. In the *Scientific Value of Tradition*, pp. 173-4, I find that I have given four instances of a very similar tradition—in Central Africa, and among "the Indian tribes of Guiana" (Brett, p. 378) and in Hayti. The first two are from letters written by Mr. H. M. Stanley from Ugigi, where the origin of Lake Tanganika was thus accounted for : A man and woman lived in possession of the secret of a fountain which contained an abundance of fish. She betrayed it to her lover, "who gazed on the brilliant creatures with admiration ; then, seized with a desire to handle one of them, he put his hand within the water . . . . when suddenly the well *burst forth*, the earth opened her womb, and soon an enormous lake replaced the plain."

Mr. Lang continues :

"The old men point spears at the boys' eyes, saying, 'If you tell this to any woman, you will die ; you will see the ground broken up and *like the sea.*' . . . As in Athens, in Syria, and among the Mandans, the Deluge tradition of Australia is connected with the mysteries. In Gippsland there is a tradition of the Deluge : 'Some children of the Kurnai, in playing about, found a *turndun,* which they took home to the camp, and showed the women. Immediately the earth crumbled away, and *it was all water*, and the Kurnai were drowned.'"

The evidence also regarding the "initiations" is scarcely less important. At least it has an equally direct bearing on our argument. As the women were more excluded from the ceremonies among the Australians than among the Mandans, although it will be remembered that their exclusion was the leading feature in the opening scenes of the Mandan ceremony, we are not surprised when we hear that "the Australian women were much less instructed in their theology than the men." Still, in the following extract from a conversation with one of their women we may see evidence of the connection of the "initiations" with the ceremony of which the noise of the "turndun" formed part. "One woman believed she had heard Pundjal, the chief supernatural being, descend in a mighty rushing noise—that is, the sound of the 'turndun' when boys were being ' *made men,*' or initiated " (Lang, p. 35).

Mr. Lang also says (p. 40): "Mr. Winwood Reade, *Savage Africa* [Captain Smith also mentions the custom in his work on *Virginia,* pp. 245-8] , reports the evidence of Mongilomba. When initiated Mongilomba was severely flogged in 'the fetich house' (as the young Spartans were flogged before the animated image of Artemis), and then he was plastered over with goat-dung" (compare *sup.* p. 47).. "Similar daubings were performed at the mysteries by the Mandans, as described · by Catlin."

"On the Congo Mr. Johnson found precisely the same ritual in the initiations " (Lang, p. 40).

Catlin (*O-kee-pa,* pp. 28-9) mentions after the scene of torture in the initiation of the young men instances of the voluntary sacrifice of several of "the little fingers of the left hand," or of "the forefinger," as "an offering to the Great Spirit," which was struck off with a hatchet on a buffalo skull.

It appears to me that this may supply a link to connect another curious description of an Australian initiation —the earliest on record with the tradition.

I find this description in David Collins's *Account of New South Wales*, 1798. The ceremony of initiation was witnessed 25th January 1795, at which the males, between the ages of eight and sixteen, "receive the qualifications which are given to them, by losing one of the front teeth" (p. 563). At the time when Mr. Collins (Judge Advocate and Secretary of the Colony) witnessed it, the tooth was ostensibly extracted as a tribute, which was exacted by the most powerful tribe of those parts from the subordinate tribes. This explanation, however, did not satisfy Mr. Collins. He noticed that the front tooth was equally absent from the mouths of the conquering tribe, and after further research he found that the teeth were thus extracted with much form and ceremony at a solemn gathering of the tribe.

Now, if we consider the matter a little, it will be seen that the demand of a front tooth would have been felt as a most wanton and tyrannous exaction even in barbarous times and by a powerful tribe. But if in its origin this abstraction of the tooth was a voluntary offering or a customary sacrifice, it would have been the tribute that would naturally have been seized upon by every conqueror, as it would have given him a ready and certain mode of ascertaining, and merely according to their customs, the adult strength of the populations subject to him.

On the other hand, from the time that the custom became associated with the notion of a tribute, its religious significance would gradually have died out.

There are, however, certain evidences which identify it more directly with the Mandan ceremony. The time of the Australian custom was when certain shrubs were flowering. The place selected "was of an oval figure, the dimensions twenty-seven feet by eighteen feet, and was named yoo-lahng." "Among them we observed *one* man *painted white* to the middle, his beard and eyebrows excepted, and altogether a frightful object" (Collins, p. 564). Compare this with the description by Catlin of the personage, " the first or only man," who opened the Mandan ceremony. " He was in appearance a very aged man . . . . with a robe. of four *white* wolves' skins. His body and face and hair were entirely covered with *white* clay, and he closely resembled, at a little distance, a *centenarian white* man " (p. 11).

The first engraving (Collins) represents the young men upon their hands and feet, imitating. the *dogs* of ' the country. In the Mandan custom, " on the entry of the ' white man,' the first order given is to muzzle all the *dogs*."

Canon Rawlinson, in his *Illustrations of the Old Testament,* p. 18, tells us that " the Cherokee Indians had a legend of the destruction of mankind by a Deluge, and of the preservation of a single family in a boat, to the construction of which they had been *incited by a dog*."

The second engraving (Collins) represents the young men seated on the ground facing a log of wood, which may have done duty for "the big canoe" which, Catlin tells us, was shaped like a "hogshead." In the engraving a man is seen carrying in " a kangaroo made of grass," concerning which presently, another carries a load of brushwood. The latter " had one or two flowering shrubs " in his nostrils.

It will be recollected that the Mandan ceremony took place when the willow first flowered.

In the third engraving we have a kangaroo dance. Now, if the kangaroo was their principal food, it would naturally supply the place of the buffalo in the Mandan custom. In the Mandan ceremony we have a buffalo-bull dance, " to the strict observance of which they attributed the coming of buffaloes to supply them with food " (Catlin). In Collins the dancers are represented advancing "in Indian file," a phrase which Catlin uses with reference to the Mandan young men just previously to the bull-dance, which occurs in corresponding sequence.

Engravings four, five, and six have only slight resemblances, and are merely preparatory to the tooth-extracting or initiation scene. There is one noticeable feature, however, that at one point the youths " uttered a mournful dismal sound like very distant thunder," suggesting the " bull-roarer." This scene is called "Boo-roo-moo-roong," a word which has resemblance in sound to boumarang, which again has resemblance in its mode of use to the " bull-roarer." In the sixth there is a presentation of spears, recalling in something the presenting the spears to the boys' eyes which Mr. Lang mentions.

Seventh scene is the extraction of the front teeth—a rough operation with a splint of wood and a piece of stone —(yet " on showing it to our medical men, they all declared they could not have been better extracted if the proper instruments had been used "). The operation was performed with minute attention to mystic rites ; the assistants all the time " made a most hideous noise in the ears of the patients, crying ' Ewah, ewah ! Gaga, gaga !' " (p. 580.) Has not this cry of " Ewah, ewah !" a resemblance to the " Evoe, evoe !" of the Bacchanals ? (Refer back to p. 47.) This scene of initiation in its leading features corresponds to the Mandan tortures and amputations.

In the eighth scene the custom closes like the Mandan in a saturnalia of riot. " Suddenly, on a signal given " (they are now seen seated on the log of wood or "big canoe " (?) which previous to initiation they had faced, sitting on the ground), " they all started up and rushed into the town, driving the men and women before them. They were now received into the class of men, were privileged to wield the spear and club. They might now also seize such females as they chose for wives."

Certain other features in common may be noticed. Previous to the tortures, the Mandan young men "lay reclining round the medicine (mystery) lodge, and had now reached the middle of the fourth day without eating, drinking, or sleeping;" and the Australians "were seated at the upper end of the yoo-lahng, each holding down his head, his hands clasped, and his legs crossed under him. In this position, awkward and painful as it must have been, we understood they were to remain all night; and, in short, until the ceremony was concluded, they were neither to look up nor take any refreshment whatsoever."

Among the Mandans, on the occasion of the festival, "an edged tool" was exacted "from every wigwam as a sacrifice *to the water*." And among the Australians, when the front teeth escaped the tribute, we catch a glimpse of the original notion. Collins says their informant told them "his own tooth was buried in the ground, and that others were thrown *into the sea*."

The teeth were apparently (p. 594) extracted by their priests, or car-rah-dis, as the fingers were lopped off among the Mandans by their priests or "medicine (mystery) men."

Captain J. G. Bourke has recently printed a valuable and interesting account[*] of the snake-dance among the Moqui Indians ; but it does not appear to have occurred to him that this dance, if not identical, is cognate with the Mandan and other diluvian traditions. This appears manifest to me from the similarity in the symbols, and in the general correspondence, so to speak, in the programme of the dance and the common fundamental notion attaching to its observance. It may, indeed, be that the Moqui Indians have not retained the tradition in the same

---

[*] *The Snake Dance of the Moquis of Arizona.* By Captain J. G. Bourke, U.S. Cavalry. (London : Sampson Low. 1884)

direct form as the Mandans. And there is this further difference, which may also account for the diluvian tradition being in their case subsidiary.

In tracing the resemblances between the Grecian festivals `certain connections were noticed between the festivals in honour of Ceres and Bacchus. But the former probably had their origin in the offering of " first-fruits," to which Aristotle refers (vide *supra*, p. 37),* and blended with the diluvian commemorative ceremonial at later date.

The snake-dance of the Moquis, as will be seen from the extracts given, has many features in common with the worship of Ceres, and was held to propitiate, if no longer the Deity, yet some higher power, to secure the growth of their corn, just as the Mandans attributed the supply of buffaloes to the fidelity with which they performed their annual " custom ;" *e.g.* (p. 161), Captain Bourke says :

" The first division in the dance remained in place, while the second, two by two, arm in arm, slowly pranced round the sacred rock, going through the motions of planting corn to the music of the monotonous dirge chanted by the first division."

And p. 123 :

" One of the old men held up a gourd-*rattle*, shook it, lifted his

* Mr. Lang (*Custom and Myth*, p. 36) suggests that the " mystica vannus Iacchi" was a mode of .raising a sacred wind, analogous to that employed by the whirlers of the " turndun." Mr. Lang, however, gives the explanation of Servius, "the ancient commentator on Virgil," who offers other explanations, among them that the " vannus " was a crate to hold offerings, "*primatias* frugum." He also gives the views of Père Lafitau, who was greatly struck with the resemblance between Greek and Iroquois or Carib initiations. He takes Servius's other explanation of the " mystica vannus," an osier vessel containing rural offerings of *first-fruits*. " This exactly answers," says Lafitau, " to that Carib ‡ Matoutou ' on which they offer sacred cassava cakes."

There is a very well-authenticated tradition of the connection of the cassava-tree with the diluvian tradition in Mr. Everard F.im Thurn's book on the *Indians of Guiana*, and among a Carib tribe, but it would run this note to too great length.

It must be borne in mind that 'Iacchus was the solemn name of Bacchus in the Eleusinian mysteries, and that during their celebration the initiated carried mystic baskets.

F

hands in an attitude of prayer towards the sun, bent down his head, moved his lips, threw his hands with fingers opened downwards towards the earth, grumbled to represent thunder, and hissed in imitation of lightning, at the same time making a sinuous line in the air with the right index-finger ; and then, seeing that my attention was fixed upon him, made a sign as if something was coming up out of the ground, and said in Spanish, ' Mucho' maiz ' (Plenty of corn), and in his own tongue, ' Polamai ' (Good)."

And p. 164 :

" The corn-meal had a sacred significance, which it might be well to bear in mind in order thoroughly to appreciate the religious import of this drama. Every time the squaws scattered. it, their lips would be detected moving in prayer."

It may, perhaps, render Captain Bourke more placable (*vide* pp. 169-70) towards these benighted people, if he realises his own evidence that the snakes merely represent the lightning and the storm, and that, therefore, the introduction of the snake into their ceremonial may not necessarily suggest demoniac associations. In the passage just cited this is evident ; but also at p. 124 it is said :

" Here also was a ground-altar. . . . The design, however, was different, and represented a bank of four layers of yellow; green, red, and white clouds, from which darted four snakes or streaks of lightning, coloured white, red, green, and yellow respectively. See Plate 19."

And it is to Captain Bourke that we are indebted for this new light. In ch. xx. he gives quotations from several authors who have written with reference to serpent-worship, and he is justified in saying (p. 225) :

" In quoting these authorities I desire to make one comment only. Not one of them has alluded to the resemblance between the undulatory motion of the serpent and the sinuous meandering of lightning, a resemblance patent to every one and portrayed by the Moquis on the altar figured on Plate 6.

It may, perhaps, be suggested that a sufficient explanation would be that lightning is frequently followed by rain, and that in Central America rain is specially invoked for the growth of corn. If, however, the lightning is the concomitant of rain, it is more generally the concomitant of the wind, the storm, the flood, and the deluge, which, if invoked, would as often have brought destruction as benefit to the crops. We might suppose, therefore, that something in milder similitude might have been found for the fertilising influence "which droppeth as the gentle rain from heaven." The present Moqui Indians, who do not appear to have retained their tradition very tenaciously, may, indeed, connect their ceremony with the invocation for rain; but that it was not the original and leading idea will, I think, be apparent when I show that the snake, and more especially the rattlesnake, the prominent snake in the Moqui ceremonial, is common to the Mandan custom, where the diluvian tradition is the predominant idea.

Catlin (p. 19) says that among the spectators of "the *bull*-dance" "there were two men called *rattlesnakes*, their bodies naked and curiously painted, resembling that reptile, each holding a *rattle* in one hand and a bunch of wild sage in the other," who, at the close of the *bull*-dance, "shook their rattles." The rattle* corresponds to "the *bull*-roarer" or "turndun," which, as Mr. Lang tells us, by an independent line of testimony, is connected with the diluvian tradition.

But curiously what I may term the counterpart evidence may be found in the names of the principal snakes selected for the mysteries. Five are mentioned by Captain Bourke, three of which, at any rate, bear in their names allusions to the Mandan symbols:

."1. 'Chúna' (*rattler*); 2. ' Le-lu-can-ga (this has yellow and black spots, and may be the *bull* snake); . . . ' Pa-chu-a ' (a *water* snake) . . ." (p. 116).

---

* Mr. Lang (p. 39) identifies the "bull-roarer" with the Grecian ρόμβος, which is ' sometimes interpreted as a magic wheel. '

I have said that the general outline or programme of the ceremonies corresponds. In the Moqui, as in the Mandan, the medicine or mystery lodge is the pivot on which the ceremony hinges; and as another instance of the interlinking of the evidence, I may mention that the mystery lodge is covered with a buffalo-skin. The prominence of the buffalo in the Mandan ceremony—the buffalo-dance, the buffalo skulls—will not have been forgotten; and here among the Moquis, in a country where the buffalo is not known, and where it must have been procured—as, in fact, Captain Bourke tells us it was—from a great distance, we find the buffalo-skin evidently used. traditionally as a central object in the snake-dance.

In the Mandan lodge the most significant symbol was the sack, or rather sacks, of water, which the Mandans pretended contained the waters of the Deluge; and among the Moquis (p. 138):

"Before the altar the Indians had now placed an earthenware bowl. . . . It was filled with *water*, and contained three large seashells. The water had a saline taste, and evidently contained 'medicine' (mystery)."

And p. 143:

"One of the old men touched the string of olivette-shells on his wrist, thus indicating a great distance in the west, . . . an explanation which I took to signify that these shells had been brought from a great distance from *the sea.* The olivette may not be a marine shell, it may be fluvial. Be that as it may, I cannot dispel from my mind a conviction that the Moquis betray, in the shells, the salt-water, sand, alabones, and other features of their dances, a derivation from a people who once knew, and perhaps worshipped, the ocean."

Before comparing the dance itself, I may again notice the rattles, which are as conspicuous here as in the Mandan ceremony, and form a connecting link with the diluvian customs elsewhere through the "bull-roarer." Mr. Lang, indeed, mentions the existence of the "bull-roarer " among the Zunis, who adjoin the Moquis; and in the Moqui dance there is mention of the "sling-whirler" (pp. 159 and 161), of "an old man who rapidly twirled a wooden sling, which emitted the shrill rumble of falling rain, so plainly heard, as the head of the procession was emerging from the arcade."

This brings us to the dance itself (ch. xv.). In the Mandan and the Australian ceremonies we noticed the prominence of an old man. The Moqui dance opens with three old men; but one seemed to act as the "head-priest," or master of ceremonies, "who stood in front of and facing the mystery lodge, holding well before him the platter of water and the eagle-feather wand," which recalls one point of the Mandan ceremonial. This old man heads the procession—"barefooted, crowned with a garland of cotton leaves, holding in his hands in front of him a flat earthen bowl, from which he sprinkled water on the ground." The second old man "carried a flat basket of fine corn-meal;" the third "rattled a fan-shaped instrument painted white." Five men and eight little boys "marched in single file," as in the Mandan and Australian ceremony, "with the same odd-looking rattles." The dancers all wore "collars of white sea-shell beads " and " alabone shells;" "the men and boys shook their rattles gently, making the music of pattering showers." "Each division marched solemnly round the sacred lodge and trees, the first division completing this formula shortly before the second." The old man in front of the lodge "began to pray in a well-modulated voice, and sprinkled the ground in front of him with more water. The first division remained in place, while the second slowly pranced around the sacred rock, going through the motions of planting corn to the music of a monotonous dirge." This ended the first act, which has resemblances to the Mandan. In the second we have the snake-dance, which corresponds to the bull-dance among the Mandans. The third ends as in the Mandan and Australian—with a general dispersal. "The Indians then grasped the snakes convulsively in great handfuls, and ran with might and main to the eastern crest of the precipice, and then darted down the trails leading to the foot, where they released the reptiles to the four quarters of the globe." "The *old man* armed *with the sling*" (whilst they were running) "twirling it vigorously, causing it to emit the same peculiar sound of *rain driven* by *the wind* which had been heard on their approach."

There are two very striking interludes in the course of the dance, to which particular attention must be drawn.

In the first, although the circumstances and surroundings are different, yet in its essential idea we seem to detect the power and victory of the woman over the serpent (Gen. iii. 15), of which we have seen evidences in the Mandan. The first division of the Moquis "remained aligned upon the sacred rock"* with the head-priest; another section approached, "their faces painted black, as with a mask of charcoal, from brow to upper lip, where the ghastly white of kaolin began . . . the crowning point being the deadly reptiles borne in the mouth"†—a picture which recalls the advent of the demon in the Mandan ceremony.

But then "the women scattering the corn-meal now developed their line more fully, . . . the main body massing between the sacred rock and sacred lodge"

---

* The "sacred rock" in the Moqui stands in the place of the "big canoe" in the Mandan ceremony. A clue to this may be discovered at p. 126 in Captain Bourke, when describing what he terms their idols, "these were *water*-worn fragments of sandstone; *water*-worn rocks roughly shaped to some sacred configuration." And if the reader will turn to Plate 2 he will find that the representation there of the sacred rock exactly corresponds to this description.

† Catlin thus describes the Mandan demon (p. 22) : "His body was painted jet-black with pulverised charcoal and grease, with rings of white clay over his limbs and body. Indentations of white, like huge teeth, surrounded his mouth, and white rings surrounded his eyes."

(comp. Mandan *supra*, p. 42). "Nearly all carried the beautiful, close-woven, flat baskets, in red, yellow, and black, ornamented with the butterfly, thunder-bird, or deer . . . from which was scattered the finely-ground corn-flour, not, as previously, on the ground, but in the air, and *upon the reptiles* as fast as thrown down. The corn-meal had a sacred significance. . . The use of the sacred meal closely resembles the crithomancy of the ancient Greeks, but is not identical with it."* At a later stage it is said the "maidens and matrons redoubled their energy, sprinkling meal not only upon the serpents wriggling at their feet, but *throwing handfuls* into the faces *of the men carrying them*" (p. 166), which forcibly reminds us of the final discomfiture of the demon in the Mandan mystery.

Taking these facts, it would seem that the introduction of the serpent into their ceremonial was only secondary and subsidiary, and not serpent-worship in the strict sense of the word. I do not recollect that there is direct evidence of it at all.

There is another very curious fact mentioned by Captain Bourke, that when the men and boys of the first division shook their rattles in the dance, the men of the second division waved their eagle feathers, all singing a refrain, "*Oh-ya-haw, oh-ya-haw*, &c.," chanted with a slow measure and graceful cadence. Compare this with the completely independent testimony as above (p. 63) of the Australian refrain in their ceremony of "*Ewah, ewah*, &c.," and the evidence (*supra*, p. 47) of *two* other similar invocations, all having a strange resemblance to the

* At page 165 it is said: "After a snake had been properly sprinkled it was picked up, generally by one of the eagle wand-bearers, but never by a woman, and carried up to the Indians of the first division," where was also the head-priest, just as the women in the Mandan ceremony assembled round the mystic man in the first instance. As, however, the serpents were specially guarded, and afterwards safely released, the superstition may, perhaps, have been that the touch of the woman might have been injurious or fatal to the serpent.

"Evoe, evoe," which St. Clement of Alexandria tells us was the cry of the Bacchanals. Compare also the "Euœ Bacche fremens" of Virgil, *Æneid*, vii. 389, ed. Heyne.

I have also come on unsuspected testimony in a most unexceptional quarter—in the pages of Mr. Everard F. im Thurn, who is ever watchfully on his guard against the action of missionaries on myths.   Compare the following extracts with the evidence already adduced.   Speaking of the dances among the Indians of Guiana, he says (p. 323): "Some beat time with hollow bamboos covered at one end with skin like a *drum*."   Some had "*whirl*-sticks, to which are tied bunches of certain seeds, which when struck against the ground clash and rattle."

"All form a procession, and march slowly round the Paiwari or liquor-trough [comp. *supra*, pp. 46, 47], droning out a chant, keeping step and waving their instruments in slow measured time. Round and round the trough the strange procession winds, all feet stamping in time the monotonous chant of *Hia-hia-hia*.   Suddenly the chant gives place to loud discordant cries, and the procession breaks up."

As regards the connection of the initiations with these feasts, there is this evidence : "In one of their Paiwari feasts there is a strange and painful dance."   "They lash each other . . . . till the blood flows freely " (comp. *sup.* p. 45), "probably originally devised as a means of testing endurance."

And Mr. Thurn says (p. 319) :

"The festivals, dances, and games originally peculiar to any people often remain but little altered long after most other matters which distinguished that people from the rest of the world have disappeared."

The initiations do not occupy so conspicuous a place in the Moqui festival, and they are in milder form.   The boy race, where the boys figure as "antelopes" in the Mandan, seems to have taken the place of the tortures, and developed into a severe trial of stamina and endurance—(" the antelope " is also one of the devices on the Moqui walls). It takes place before the dance. " Every one of the men was streaming with perspiration, and the thumping of hearts and wheezing of lungs could be plainly heard." A further form of initiation appears during the dance in the handling of the serpents by the youths. " An infant Hercules " stoutly and bravely upheld a five-foot monster.

There is, however, a remarkable coincidence.

At p. 133, Captain Bourke tells us that Tochi, their guide, told him that when the rain did not come from the sky, the Moquis came into this "estufa" and "danced for it ;" that here also came the young men to be baptised for medicine (mystery) men—"bautista por cochinos." The guide spoke in broken Spanish. Captain Bourke adds :

"I made Tochi repeat all he had said, and then asked for an explanation. He said that, after *all* the big dances—as, for ex-ample, after *that* of to-day—the young men who were to learn all the secrets would come to one of the estufas, and there have their heads washed with water by the old men. As he said this, he made the motion of pouring a few drops of water upon the head of some one kneeling beneath him."

Captain Bourke remained in doubt how far "this lus-tration partook of the nature of the sacrament of baptism," and how far it was " likely to be a reminiscence of the teachings of the early missionaries . . . ." Now for the coincidence. If it was the reminiscence of baptism rather than of the prefigurement of baptism, how are we to account for the apparently identical initiations among the Maoris of New Zealand?

I find in *The Natural History of Man*, by the Rev. J. G. Wood, M.A. (Routledge, 1870), p. 177 :

"Youths undergo a long course of instruction before they can take rank among the priests. Dr. Dieffenbach was fortunate enough to witness a portion of this instruction. ' I was present at one of the lessons. An old priest was sitting under a tree, and at his feet was a boy, his relative. He listened attentively to the repetition of certain words, which seemed to have no meaning, but which it must have required a good memory to retain in their due order. At the old Tohunga's side was part of a man's skull filled with water; but from time to time he *dipped a green branch* which he moved over the boy's head. . . .' "

The resemblance in the external forms of the initiation will be noticed, and beyond it the symbols of the water and the green branches, which a slight reference to the preceding pages will show to have been intimately associated with the Mandan and Moqui rites.

One thing is obscure : how it has come about that the Moquis of Arizona—the *Snake* Indians—call themselves Opii. "The Moquis call themselves Hopii, or Opii, a term *not now* in the language of every-day life, but referring in some way to the Pueblo custom of banging the hair at the level of the eyebrows . . . ." (Captain Boúrke, p. 117). This surmise, however, leaves untouched this difficulty, that the name Opii is so suggestive of ophiolatry ; and although, as I have contended, the Snake Indians are not in any strict sense serpent-worshippers, yet the affinity of their name with the Greek Oφίς is, to say the least, striking and surprising. I do not venture, however, to attempt to solve the difficulty, but must leave it as a " crux " to the philologers.

Additional evidence will be found in chapter xi. (with Appendix) of *Tradition with Reference to Mythology* (Burns & Oates, 1872) in respect to the Mandan, Dahoman, Pongol (India), Pota (India), p. 247, and Patagonian festivals.

CHAPTER V.

THERE still remain passages in Plato's discourse which have to be cleared up before the ground can be said to be exhausted. There is one statement especially which the Egyptian priests of Sais made to Solon, as Critias is made to say, which must not be forgotten, from Mr. Donnelly's point of view. I mean the statement in *Timæus*, p. 517 (Jowett), "about the greatest action which the Athenians ever did, and which ought to be most famous, but which, through lapse of time and the destruction of the actors, has not come down to us." And p. 521: "Many great and wonderful deeds are recorded of your state (Athens) in our histories. But one of them exceeds all the rest in greatness and valour. For these histories tell us of a mighty power which was aggressing wantonly against the whole of Europe and Asia, and to which your city put an end. This power came forth from the Atlantic Ocean. . . . ."

If Atlantis is a fiction of Plato's, every tradition imported into his discourse will naturally be located in Atlantis. As regards the mighty power which was aggressing against the whole of Europe, the view which had occurred to me I find accords with Professor Jowett's, which I will give accordingly, adding a few words in further eluci³ation. Professor Jowett says (ii. 589): "This mythical conflict is prophetic or symbolical of the struggle of Athens and Persia, and, perhaps, in some degree also of the wars of the Greeks and Carthaginians, in the same way that the Persian is prefigured by the

Trojan War to the mind of Herodotus, or as the narrative of the first part of the *Æneid* foreshadows the wars of Carthage and Rome."

That what is set down in the " sacred register " of the Egyptian priests (p. 520) cannot be taken as strictly historical is obvious from what Professor Jowett points out—that the statement that the war occurred " nine thousand years ago " is slightly inconsistent with the statement " which gives the same date for the foundation of the city " (p. 590). Professor Jowett's view of *Atlantis* (p. 589) is that " we may safely conclude that the entire narrative is due to the imagination of Plato." I have in a previous chapter hazarded the conjecture that it was founded on the narrative of Hanno.

1. I will now consider, in the first place, how far the statements in Plato's discourse that do not correspond to facts in the narrative of Hanno were imported by Plato himself, and secondly, how far they may have been transmitted through Solon.

If the statement we are now considering was a confused and exaggerated tradition of the Persian War, it must have been imported by Plato ; and I may suggest that the confusion and exaggeration may have come about in this way, and if so without being entirely due to the imagination of Plato. If Plato was himself in Egypt, as Valerius Maximus (Lewes's *Hist. of Phil.*, C. W. Collins's *Life of Plato*) tells us he was, or even during his residence at Syracuse, he might have heard the statements, such as were not derived from Hanno, which he attributes to Solon. Now, if we recollect that Plato's visit would have taken place during the Persian domination in Egypt,[*]

---

[*] The Persian rule in Egypt may be broadly said to have extended from B C. 527, when Cambyses overthrew the dynasty of Sais, till B.C. 332. The intervention of the XXIX. and XXX. Dynasties would not have affected the sentiment at Sais as against the Persians. The dates are from Brugsch's *Egypt*.

this legend of a gigantic war, in which Athens was victorious and "brought to an end " (a phrase, by the bye, applicable to the termination of the Persian War, but inconsistent with the destruction of both parties to the contest through "the subsidence of Atlantis "), is precisely the recollections which, in its distorted circumstances, may have remained in Egypt of the Græco-Persian War, and which the priests of Sais, "the city from which Amasis the .king was sprung," and where a Greek colony had been established from an early date (*circa* 660 B.C.), would have cherished and magnified in their legend as against the Persian domination, which had engulphed them, as well as the rest of Greece, in a common deluge, subsidence, and destruction, and yet which, so long as the domination lasted, they would, perhaps, deem it prudent to veil under a legendary disguise.

We do not, perhaps, sufficiently realise that history was not then digested as it is now, however inaccurately, into recognised record ; and although MSS. of Herodotus and Thucydides, and other histories, were drifting, the recollection of events was, in the main, traditional.

2. In chapter ii., when discussing Plato's *Atlantis* with reference to the report of Hanno, I omitted to advert to a digression on the laws and institutions of the island of Atlantis, as I considered that it would be better reserved until we came to the consideration how far other traditions might have been imported. Now, what we are told of these customs, which "were regulated by the injunctions of Poseidon as the law had handed them down," and "were inscribed by the *first men* on a column of orichalcum, ..." bears a close resemblance to what is recorded of the Amphictyonic council and to the Areopagus—institutions which had fallen into comparative insignificance since the time of Pericles, and whose authority Plato might not unnaturally seek indirect

means of reviving. Just as we are told that the laws of Atlantis were injunctions of Poseidon and handed down by the first men, so we find the Amphictyonic laws attributed to the son of Deucalion, and that the Ionian federation held their assemblies at a sacred place on Mount Mycale, where they had dedicated in common a temple to Neptune. Plato tells us that the laws were engraved on a column of orichalcum; and it is recorded that the terms of the Latin League—an analogous confederation—were engraved " on a brazen column," which was preserved in the time of Dionysius of Halicarnassus, a fact which might not impossibly have come at earlier date to the knowlege of Plato also. The Atlanteans " were to deliberate together about war and other matters, and were not to take up arms against one another," just as, according to their oath, referred to by Æskines " as the ancient oath of the Amphictyons," they were bound " not to destroy each other's cities or debar them from the use of their fountains in peace and war. . . ." Plato adds, " There were many special laws which the several kings had inscribed about the temple." This is, perhaps, intended to cover the introduction of certain other customs which may be identified with what has come down to us of the mode of proceeding of the Areopagus. There was a special reason why they should be imported into the narrative on the particular occasion, as the institution of the Areopagus was by some attributed to Solon and by others to Cecrops, both in the line of his reputed ancestors.

The judges of the Areopagus " always sat in the open air, because they took cognisance of murder; and by their laws it was not permitted for the murderer and his accuser to be both under the same roof. . . . They always heard causes and passed sentence in the night, that they might not be prepossessed in favour of the plaintiff or the defendant by seeing them." In the confederacy of the Atlanteans the judicial proceedings are also described as in the open air; and " when *darkness* came on, and the fire about the sacrifice was cool, all of them put on most beautiful azure robes, and sitting on the ground at night, ... if any one had any accusation to bring against any one; and *when* they *had* given judgment, at daybreak they wrote down their sentences on a golden tablet, and deposited them as memorials with their robes " (Plato's *Atlantis*, Jowett). I should have mentioned—and it should be remarked also in the almost inseparable connection of the ox or the bull with the diluvian commemorative customs—that " before the Amphictyons proceeded to business they sacrificed an ox, and cut his flesh *into small pieces*, intimating that union and unanimity prevailed in the several cities they represented." Plato says when the Atlanteans gathered together, " before they passed judgment they gave their pledges to one another in this wise: there were bulls who had the range of the temple of Poseidon, and the ten who were left alone in the temple, after they had offered prayers to the gods that they might take the sacrifices which were acceptable to them, hunted the bulls without weapons, but with staves and nooses, and the bull which they caught they led up to the column." " When, therefore, after offering sacrifice according to their customs, they had burnt the limbs of the bull, they *mingled a cup* and cast in a clot of blood for *each of them*. . . ." Although it may be objected that these extracts are taken somewhat at random from Plato's narrative, yet, on being pieced together, they will be found to exhaust all the disclosures of Plato respecting Atlantis.

We will now consider (it being conceded that the basis of Plato's fiction was the *Periplus* of Hanno, and that he has imported more recent facts, as, for instance, the incidents of the Persian War) whether something of the original figment of *Atlantis* might not have been handed down, as

Plato says it was, in family tradition from Solon, and reconstructed or adapted by Plato.

The conception of Atlantis probably originated in some development of the diluvian tradition. It is curious, however, that it might have been brought prominently to the notice and speculation of Solon or the priests of Sais under circumstances very similar to those of which we have evidence—in the case of Plato as regards Hanno, viz. through the circumnavigation of Africa during the reign of Necho or Nako (B.C. 611). No record, however, of this exploration has survived.

Even if we adopt this conjecture, yet if Plato had not seen the MS. of Solon ("My great-grandfather Dropidas had the original writing, which is still in my possession, and was carefully studied by me *when I was a child*") since he was ten years old (" I will tell an old-world story, which I heard from an aged man; for Critias was, as he said, at that time nearly ninety years of age, and I was about ten years old "), he might naturally have based or rebased the fiction on the narrative more recently within his knowledge than that which had faded from his recollection. He uses such phrases as " If I can recollect and recite enough of what was said by the priests," and " If I have not forgotten what I heard when I was a child," " and I would specially invoke Mnemosyne," which imply that, if any such MS. existed, he was separated from it, and had not seen it since he was a child.

We know little more of the life of Solon than what Plutarch has preserved, but the little that is told us describes a situation, in which the production of a fiction such as *Atlantis* in covert allusion would be exceedingly natural. We are told that

" The occasion which first brought Solon prominently forward as an actor on the political stage was the contest between Athens and Megara respecting the possession of Salamis. The ill-success of

the attempts . . . had led to the enactment of a law forbidding the
writing or saying anything to urge the Athenians to renew the contest. Solon, indignant at the dishonourable renunciation of their
claims, and seeing that many of the younger and more impetuous
citizens were only deterred by the law from proposing a fresh
attempt, . . . hit upon the device of feigning to be mad and
causing a report of his condition to be spread over the city;
whereupon he rushed into the *agora*, mounted the herald's stone,
and there recited a short elegiac poem of one hundred lines,   . .
calling upon the Athenians to retrieve their disgrace and reconquer
the lovely island."

This led to the repeal of the enactment. If, in the
interval, encouraged by his success, he had covertly sought
to excite their enthusiasm in the cause of Salamis, " the
lovely island," would not an allegory in the disguise of
Atlantis have admirably subserved his purpose? In this
case the legend must have taken a different shape from
that in which it was presented by Plato. The description,
however, in Plato would much more exactly befit the
dimensions of Salamis than those of a " lost continent,"
as we have already seen in the way in which the descriptions would equally fit the account given by Hanno of the
islands he visited in the course of his exploration. Solon's
love for Salamis would appear to have been the abiding
sentiment of his life; but the secret of it is not so
apparent.

It may possibly be accounted for in this way: Solon
is said to have been the descendant of Codrus, the last
King of Athens, and Codrus was the reputed representative
of Cecrops. Cecrops, however, is enveloped in the mists
of a mythical age. He undoubtedly represents a local
aboriginal ancestor, but at the same time we find him
invested with the features and attributes of the primeval
progenitor : like Oannes and Dagon, he is half-man, half-
fish ; like Bacchus and Saturn, he is the first cultivator.
" The different mythical personages of this name in Bœotia

and Eubœa are only multiplications of the one original hero " (Smith, *Myth. Dict.*). Dr. Smith says, with reference to the statement that he migrated from Sais in Egypt, " But this account is not only rejected by some of the ancients themselves, but by the ablest critics of modern times " (Müller, Thirlwall).

There is a personage of this mythical age who may possibly be regarded as in one degree less mythical than Cecrops—Cenchreus or Cychreus, King of Salamis. If Cecrops is represented as half-man, half-dragon, Cychreus is said to have delivered Salamis from a dragon. He is fabled to have been the son of Salamis, who gave her name to the island by Poseidon, and whose mother's name was Asopis. Cychreus-Asopis might pass in contraction into Cecrops, and as King of Salamis might have extended his dominion to the mainland. If he was the Cecrops who founded the Athenian dynasty, and if the dynasty was thus associated with Salamis, it would account for the predilection and love of his descendant Solon for " the lovely island." The sentiment of Solon almost renders· such an origin probable, and there is a slight confirmation of this connection between Athens and Salamis in the later legend, that while the battle of Salamis was going on, a dragon appeared in one of the Athenian ships, and that an oracle declared this dragon to be Cychreus (Smith, *Myth. Dict.*).

This speculation concerning Solon would seem to require that he should have travelled in Egypt in early life, and Plutarch (Smith, *Myth. Dict.*) appears to be the authority for the statement that in his youth he sought his fortune as a foreign trader.

Plato says that Solon derived his information from the priests of Sais, " the city from which Amasis the king was sprung." If this is intended to mean that he visited Sais when Amasis (Aahmes) was king, he must have visited Egypt in his latter years : Solon, *circa* B.C. 638-520 (Smith, *Myth. Dict.*) ; Aahmes or Amasis, 572 B.C. (Brugsch) ; 569-525 B.C. (Lenormant). If, however, his dying instructions were that his ashes were to be scattered on the soil of Salamis—and, at any rate, the tradition is recorded by Aristotle (Smith, *Dict.*)—his sentiment regarding the island would appear to have been as strong in his old age as in his youth, and similar reasons for his covert allusions to Salamis might have existed then.

This exhausts the few facts which were available for the inquiry so far as Solon is concerned. In venturing the theory as to the Periplus of Hanno, we seemed to touch more tangible evidence, and to stand on firmer ground. Beyond this point we can only float in a " mare di sargasso " of conjecture.

But, however the exigencies of historical truth may compel us to discard the legend of Atlantis from the ground of history, few would wish to see it banished from the regions of poetry and imagination ; and here we must recur to the pages of Mr. Donnelly, and express the hope that if the legend should die out everywhere else, it may survive in the charming lines which I append :*

> " ' Mother, I've been on the cliffs out yonder,
>   Straining my eyes o'er the breakers free
> To the lovely spot where the sun was setting,
>   Setting and sinking into the sea.
>
> The sky was full of the fairest colours—
>   Pink and purple and paly green,
> With great soft masses of gray and amber,
>   And great bright rifts of gold between.

* Mr. Donnelly's *Atlantis*, p. 421. These extracts are given "from a poem of Miss Eleanor C. Donnelly of Philadelphia, ' The Sleeper's Sail,' where the starving boy dreams of the pleasant and plentiful land." The lines follow as above.

And all the birds that way were flying,
  Heron and curlew overhead,
With a mighty eagle westward floating,
  Every plume in their pinions red.

And then I saw it, the fairy city,
  Far away o'er the waters deep;
Towers and castles and chapels glowing,
  Like blessèd dreams that we see in sleep.

What is its name?'   ' Be still, *acushla*
  (Thy hair is wet with the mists, my boy);
Thou hast looked perchance on the Tir-na-n'oge,
  Land of eternal youth and joy!

Out of the sea, when the sun is setting,
  · It rises golden and fair to view ;
No trace of ruin, or change of sorrow,
  No sign of age where all is new.' . . .

The starving child seeks to reach this blessed land in a boat,
and is drowned.

High on the cliffs, the lighthouse-keeper
  Caught the sound of a piercing scream ;
Low in her hut, the lonely widow
  Moaned in the maze of a troubled dream ;

And saw in her sleep a seaman ghostly,
  With seaweeds clinging in his hair,
Into her room, all wet and dripping,
  A drownèd boy on his bosom bear.

Over Death Sea, on a bridge of silver,
  The child to his Father's arms had passed ;
Heaven was nearer than Tir-na-n'oge,
  And the golden city was reached at last."

APPENDIX A.

The Periplus or Voyage of Hanno, commander of the Carthaginians, B.C. 515 (Lenormant), from Heeren's *Historical Researches: Africa*, p. 478. Heeren says : " I cannot, however, believe that any critic will in the present day doubt its authenticity in the whole, though they may its completeness." M. F. Lenormant expresses no doubt. He says : " The official report of the voyage of Hanno round the coast of Africa, deposited in the temple of Baal-Hamon [in the Greek text " of Kronos "*] at Carthage, has been preserved to us in its entirety in a Greek version." He adds, " that it is the single historical Carthaginian document of any extent which has reached us " (*Hist. Anc.* ii. p. 413).

" It was decreed by the Carthaginians that Hanno should undertake a voyage beyond the Pillars of Hercules, and found Liby-Phœnician cities. [The colonists which Hanno carried out consisted, as we are expressly informed, of Liby-Phœnicians, and were not chosen from among the citizens of Carthage, but taken from the country inhabitants.] He sailed accordingly with sixty ships of fifty oars each, and a body of men and women to the number of 30,000, and provisions and other necessaries.

" When we had passed the Pillars on our voyage, and had sailed beyond them for two days, we founded the first city, which we named Thymiaterium. Below it lay an extensive plain. Proceeding thence towards the west, we

* Vide *supra*, p. 23.

came to Solocis, a promontory of Libya, a place thickly
covered with trees, where we erected a temple to Neptune;
and again proceeded for the space of half a day towards
the east, until we arrived at a lake lying not far from the
sea and filled with abundance of large reeds. Here elephants and a great number of other wild beasts were
feeding.

"Having passed the lake about a day's sail, we founded
cities near the sea called Cariconticos and Gytte and Acra
and Melitta and Arambys. Thence we came to the great
river Lixus, which flows from Libya. On its banks the
Lixitæ, a shepherd tribe, were feeding flocks, amongst
whom we continued some time on friendly terms. Beyond
the Lixitæ dwelt the inhospitable Ethiopians, who pasture
a wild country intersected by large mountains, from which
they say the river Lixus flows. In the neighbourhood of
the mountains lived the Troglodytæ, men of various appear-
ances, whom the Lixitæ described as swifter in running
than horses.

"Having procured interpreters from them, we coasted
along a desert country towards the south two days. Thence
we proceeded towards the east the course of a day. Here
we found in a recess of a certain bay a small island containing a circle of five stadia, where we settled a colony
and called it Cerne. We judged from our voyage that this
place lay in a direct line with Carthage; for the length of
our voyage from Carthage to the Pillars was equal to that
from the Pillars to Cerne.

"We then came to a lake, which we reached by sailing
up a large river called Chretes. This lake had three
islands larger than Cerne; from which, proceeding a day's
sail, we came to the extremity of the lake, which was
overhung by large mountains, inhabited by savage men
clothed in skins of wild beasts, who drove us away by
throwing stones and hindered us from landing. Sailing

thence, we came to another river that was large and broad and full of crocodiles and river-horses; whence, returning back, we came again to Cerne.

"Thence we sailed towards the south twelve days, coasting the shore, the whole of which is inhabited by Ethiopians, who would not wait our approach, but fled from us. Their language was not intelligible even to the Lixitæ who were with us. Towards the last day we approached some large mountains covered with trees, the wood of which was sweet-scented and variegated. Having sailed by these mountains for two days, we came to an immense opening of the sea, on each side of which, towards the continent, was a plain, from which we saw by night fire arising at intervals in all directions, more or less.

"Having taken in water there, we sailed forward five days near the land, until we came to a large bay, which, our interpreters informed us, was called Western Horn. In this was a large island, and in the island a salt-water lake, and in this another island, where when we landed we could discover nothing in the day-time except trees, but in the night we saw many fires burning, and heard the sound of pipes, cymbals, drums, and confused shouts. We were then afraid, and our diviners ordered us to abandon the island. Sailing quickly away thence, we passed a country burning with fires and perfumes; and streams of fire supplied from it fell into the sea. The country was impassable on account of the heat. We sailed quickly thence, being much terrified; and passing on for four days, we discovered a country full of fire. In the middle was a lofty fire, larger than the rest, which seemed to touch the stars. When day came, we discovered it to be a large hill, called the Chariot of the Gods. On the third day after our departure thence, having sailed by those streams of fire, we arrived at a bay called the Southern

Horn, at the bottom of which lay an island like the former, having a lake, and in this lake another island, full of savage people, the greater part of whom were women, whose bodies were hairy, and whom our interpreters called Gorillæ. Though we pursued the men, we could not seize any of them; but all fled from us, escaping over the precipices and defending themselves with stones. Three women were, however, taken; but they attacked their conductors with their teeth and hands, and could not be prevailed upon to accompany us. Having killed them, we flayed them and brought their skins with us to Carthage. We did not sail further on, our provisions failing us."

APPENDIX B.

PLATO'S ATLANTIS.

[Professor B. Jowett's *Plato*, ii. pp. 602-612; *Critias*, 113-119.]

" THE tale, which was of great length, began as follows:
I have before remarked, in speaking of the allotments of
the gods, that they distributed the whole earth into portions differing in extent, and made themselves temples
and sacrifices. And Poseidon, receiving for his lot the
island of Atlantis, begat children by a mortal woman, and
settled them in a part of the island which I will proceed
to describe. On the side toward the sea, and in the centre
of the whole island, there was a plain which is said to
have been the fairest of all plains, and very fertile. Near
the plain again, and also in the centre of the island, at a
distance of about fifty stadia, there was a mountain, not
very high on any side. In this mountain there dwelt one
of the earth-born primeval men of that country, whose
name was Evenor, and he had a wife named Leucippe,
and they had an only daughter, who was named Cleito.
The maiden was growing up to womanhood when her
father and mother died; Poseidon fell in love with her,
and had intercourse with her; and, breaking the ground,
enclosed the hill in which she dwelt all round, making
alternate zones of sea and land, larger and smaller, encircling one another; there were two of land and three of
water, which he turned as with a lathe out of the centre
of the island, equidistant every way, so that no man could
get to the island, for ships and voyages were not yet heard
of. He himself, as he was a god, found no difficulty in
making special arrangements for the centre island, bring-

ing two streams of water under the earth, which he caused
to ascend as springs, one of warm water and the other of
cold, and making every variety of food to spring up abundantly in the earth. He also begat and brought up five
pairs of male children, dividing the island of Atlantis into
ten portions : he gave to the first-born of the eldest pair
his mother's dwelling and the surrounding allotment,
which was the largest and best, and made him king over
the rest ; the others he made princes, and gave them rule
over many men and a large territory. And he named
them all : the eldest, who was king, he named Atlas, and
from him the whole island and the ocean received the
name of Atlantic. To his twin-brother, who was born
after him, and obtained as his lot the extremity of the
island towards the Pillars of Heracles, as far as the
country which is still called the region of Gades in that
part of the world, he gave the name which in the Hellenic
language is Eumelus, in the language of the country
which is named after him, Gadeirus. Of the second pair
of twins, he called one Ampheres and the other Evæmon.
To the third pair of twins he gave the name Mneseus to
the elder, and Autochthon to the one who followed him.
Of the fourth pair of twins, he called the elder Elasippus
and the younger Mestor. And of the fifth pair, he gave
to the elder the name of Azaes, and to the younger Dia-
prepes. All these and their descendants were the inhabitants and rulers of divers islands in the open sea ; and
also, as has been already said, they held sway in the
other direction over the country within the Pillars as far
as Egypt and Tyrrhenia. Now Atlas had a numerous and
honourable family, and his eldest branch always retained
the kingdom, which the eldest son handed on to his eldest
for many generations ; and they had such an amount of
wealth as was never before possessed by kings and poten-
tates, and is not likely ever to be again, and they were

furnished with everything which they could have, both in city and country. For, because of the greatness of their empire, many things were brought to them from foreign countries, and the island itself provided much of what was required by them for the uses of life. In the first place, they dug out of the earth whatever was to be found there, mineral as well as metal, and that which is now only a name, and was then something more than a name—orichalcum—was dug out of the earth in many parts of the island, and, with the exception of gold, was esteemed the most precious of metals among the men of those days. There was an abundance of wood for carpenters' work, and sufficient maintenance for tame and wild animals. Moreover, there were a great number of elephants in the island, and there was provision for animals of every kind, both for those which live in lakes and marshes and rivers, and also for those which live in mountains and on plains, and therefore for the animal which is the largest and most voracious of them. Also, whatever fragrant things there are in the earth, whether roots, or herbage, or woods, or distilling drops of flowers or fruits, grew and thrived in that land; and again, the cultivated fruit of the earth, both the dry edible fruit and other species of food, which we call by the general name of legumes, and the fruits having a hard rind, affording drinks, and meats, and ointments, and good store of chestnuts and the like, which may be used to play with, and are fruits which spoil with keeping—and the pleasant kinds of dessert which console us after dinner, when we are full and tired of eating—all these that sacred island lying beneath the sun brought forth fair and wondrous in infinite abundance. All these things they received from the earth, and they employed themselves in constructing their temples, and palaces, and harbours, and docks; and they arranged the whole country in the following manner: First of all they bridged

over the zones of sea which surrounded the ancient metro-
polis, and made a passage into and out of the royal palace ;
and then they began to build the palace in the habitation
of the god and of their ancestors. This they continued to
ornament in successive generations, every king surpassing
the one who came before him to the utmost of his power,
until they made the building a marvel to behold for size
and for beauty. ·And, beginning from the sea, they dug
a canal three hundred feet in width and one hundred feet
in depth, and fifty stadia in length, which they carried
through to the outermost zone, making a passage from
the sea up to this, which became a harbour, and leaving
an opening sufficient to enable the largest vessels to find
ingress. Moreover, they divided the zones of land which
parted the zones of sea, constructing bridges of such a
width as would leave a passage for a single trireme to pass
out of one into another, and roofed them over; and there
was a way underneath for the ships, for the banks of the
zones were raised considerably above the water. Now the
largest of the zones into which a passage was cut from
the sea was three stadia in breadth, and the zone of land
which came next of equal breadth; but the next two, as
well the zone of water as of land, were two stadia, and
the one which surrounded the central island was a stadium
only in width. The island in which the palace was situated had a diameter of five stadia. This, and the zones
and the bridge, which was the sixth part of a stadium in
width, they surrounded by a stone wall, on either side
placing towers, and gates on the bridges where the sea
passed in. The stone which was used in the work they
quarried from underneath the centre island and from
underneath the zones, on the outer as well as the inner
side. One kind of stone was white, another black, and a
third red; and, as they quarried, they at the same time
hollowed out docks double within, having roofs formed out

of the native rock. Some of their buildings were simple, but in others they put together different stones, which they intermingled for the sake of ornament, to be a natural source of delight. The entire circuit of the wall which went round the outermost one they covered with a coating of brass, and the circuit of the next wall they coated with tin, and the third, which encompassed the citadel, flashed with the red light of orichalcum. The palaces in the interior of the citadel were constructed in this wise: In the centre was a holy temple dedicated to Cleito and Poseidon, which remained inaccessible, and was surrounded by an enclosure of gold; this was the spot in which they originally begat the race of the ten princes, and thither they annually brought the fruits of the earth in their season from all the ten portions, and performed sacrifices to each of them. Here, too, was Poseidon's own temple, of a stadium in length and half a stadium in width, and of a proportionate height, having a sort of barbaric splendour. All the outside of the temple, with the exception of the pinnacles, they covered with silver, and the pinnacles with gold. In the interior of the temple the roof was of ivory, adorned everywhere with gold and silver and orichalcum; all the other parts of the walls and pillars and floor they lined with orichalcum. In the temple they placed statues of gold: there was the god himself standing in a chariot —the charioteer of six winged horses—and of such a size that he touched the roof of the building with his head; around him there were 'a hundred Nereids riding on dolphins, for such was thought to be the number of them in that day. There were also in the interior of the temple other images which had been dedicated by private individuals. And around the temple on the outside were placed statues of gold of all the ten kings and of their wives; and there were many other great offerings, both of kings and of private individuals, coming both from the city itself and the foreign cities over which they held sway. There was an altar, too, which in size and workmanship corresponded to the rest of the work, and there were palaces in like manner which answered to the greatness of the kingdom and the glory of the temple.

"In the next place, they used fountains both of cold and hot springs; these were very abundant, and both kinds wonderfully adapted to use by reason of the sweetness and excellence of their waters. They constructed buildings about them, and planted suitable trees; also cisterns, some open to the heaven, others which they roofed over, to be used in winter as warm baths: there were the king's baths, and the baths of private persons, which were kept apart; also separate baths for women, and others again for horses and cattle, and to them they gave as much adornment as was suitable for them. The water which ran off they carried, some to the grove of Poseidon, where were growing all manner of trees of wonderful height and beauty, owing to the excellence of the soil; the remainder was conveyed by aqueducts which passed over the bridges to the outer circles: and there were many temples built and dedicated to many gods; also gardens and places of exercise, some for men, and some set apart for horses, in both of the two islands formed by the zones; and in the centre of the larger of the two there was a racecourse of a stadium in width, and in length allowed to extend all round the island, for horses to race in. Also there were guard-houses at intervals for the body-guard, the more trusted of whom had their duties appointed to them in the lesser zone, which was nearer the Acropolis; while the most trusted of all had houses given them within the citadel, and about the persons of the kings. The docks were full of triremes and naval stores, and all things were quite ready for use. Enough of the plan of the royal palace. Crossing the outer harbours, which were three in number, you would come to a wall which began at the sea and went all round : this was everywhere distant fifty stadia from the largest zone and harbour, and enclosed the whole, meeting at the mouth of the channel toward the sea. The entire area was densely crowded with habitations; and the canal and the largest of the harbours were full of vessels and merchants coming from all parts, who, from their numbers, kept up a multitudinous sound of human voices and din of all sorts night and day. I have repeated his descriptions of the city and the parts about the ancient palace nearly as he gave them, and now I must endeavour to describe the nature and arrangement of the rest of the country. The whole country was described as being very lofty and precipitous on the side of the sea, but the country immediately about and surrounding the city was a level plain, itself surrounded by mountains which descended toward the sea ; it was smooth and even, but of an oblong shape, extending in one direction three thousand stadia, and going up the country from the sea through the centre of the island two thousand stadia ; the whole region of the island lies toward the south, and is sheltered from the north. The surrounding mountains he celebrated for their number and size and beauty, in which they exceeded all that are now to be seen anywhere ; having in them also many wealthy inhabited villages, and rivers and lakes, and meadows supplying food enough for every animal, wild or tame, and wood of various sorts, abundant for every kind of work. I will now describe the plain, which had been cultivated during many ages by many generations of kings. It was rectangular, and for the most part straight and oblong ; and what it wanted of the straight line followed the line of the circular ditch. The depth and width and length of this ditch were incredible, and gave the impression that such a work, in addition to so many other works, could hardly have been wrought by the hand of man. But I must say what I have heard. It was excavated to the depth of a hundred feet, and its breadth was a stadium everywhere; it was carried round the whole of the plain, and was ten thousand stadia in length. It received the streams which came down from the mountains, and winding round the plain, and touching the city at various points, was there let off into the sea. From above, likewise, straight canals of a hundred feet in width were cut in the plain, and again let off into the ditch, toward the sea; these canals were at intervals of a hundred stadia, and by them they brought down the wood from the mountains to the city, and conveyed the fruits of the earth in ships, cutting transverse passages from one canal into another, and to the city. Twice in the year they gathered the fruits of the earth—in winter having the benefit of the rains, and in summer introducing the water of the canals. As to the population, each of the lots in the plain had an appointed chief of men who were fit for military service, and the size of the lot was to be a square of ten stadia each way, and the total number of all the lots was sixty thousand.

" And of the inhabitants of the mountains and of the rest of the country there was also a vast multitude having leaders, to whom they were assigned according to their dwellings and villages. The leader was required to furnish for the war the sixth portion of a war-chariot, so as to make up a total of ten thousand chariots; also two horses and riders upon them, and a light chariot without a seat, accompanied by a fighting man on foot carrying a small shield, and having a charioteer mounted to guide the horses; also, he was bound to furnish two heavy-armed men, two archers, two slingers, three stone-shooters, and three javelin men, who were skirmishers, and four sailors to make up a complement of twelve hundred ships. Such was the order of war in the royal city—that of the other nine governments was different in each of them, and would be wearisome to relate. As to offices and honours, the following was the arrangement from the first: Each of the ten kings, in his own division and in his own city, had the absolute control of the citizens, and in many cases of the laws, punishing and slaying whomsoever he would.

"Now the relations of their goverments to one another were regulated by the injunctions of Poseidon as the law had handed them down. These were inscribed by the first men on a column of orichalcum, which was situated in the middle of the island, at the temple of Poseidon, whither the people were gathered together every fifth and sixth year alternately, thus giving equal honour to the odd and to the even number. And when they were gathered together they consulted about public affairs, and inquired if any one had transgressed in anything, and passed judgment on him accordingly—and before they passed judgment they gave their pledges to one another in this wise: There were bulls who had the range of the temple of Poseidon; and the ten who were left alone in the temple, after they had offered prayers to the gods that they might take the sacrifices which were acceptable to them, hunted the bulls without weapons, but with staves and nooses; and the bull which they caught they led up to the column; the victim was then struck on the head by them, and slain over the sacred inscription. Now on the column, besides the law, there was inscribed an oath invoking mighty curses on the disobedient. When, therefore, after offering sacrifice according to their customs, they had burnt the limbs of the bull, they mingled a cup and cast in a clot of blood for each of them; the rest of the victim they took to the fire, after having made a purification of the column all round. Then they drew from the cup in golden vessels, and, pouring a libation on the fire, they swore

H

that they would judge according to the laws on the column, and would punish any one who had previously transgressed, and that for the future they would not, if they could help, transgress any of the inscriptions, and would not command or obey any ruler who commanded them to act otherwise than according to the laws of their father Poseidon. This was the prayer which each of them offered up for himself and for his family, at the same time drinking, and dedicating the vessel in the temple of the god; and, after spending some necessary time at supper, when darkness came on and the fire about the sacrifice was cool, all of them put on most beautiful azure robes, and, sitting on the ground at night near the embers of the sacrifices on which they had sworn, and extinguishing all the fire about the temple, they received and gave judgment, if any of them had any accusation to bring against any one; and, when they had given judgment, at daybreak they wrote down their sentences on a golden tablet, and deposited them as memorials with their robes. There were many special laws which the several kings had inscribed about the temples, but the most important was the following: That they were not to take up arms against one another, and they were all to come to the rescue if any one in any city attempted to overthrow the royal house. Like their ancestors, they were to deliberate in common about war and other matters, giving the supremacy to the family of Atlas; and the king was not to have the power of life and death over any of his kinsmen, unless he had the assent of the majority of the ten kings.

" Such was the vast power which the god settled in the lost island of Atlantis; and this he afterwards directed against our land on the following pretext, as tradition tells: For many generations, as long as the divine nature lasted in them, they were obedient to the laws, and well-affectioned toward the gods, who were their kinsmen; for they possessed true and in every way great spirits, practising gentleness and wisdom in the various chances of life, and in their intercourse with one another. They despised everything but virtue, not caring for their present state of life, and thinking lightly on the possession of gold and other property, which seemed only a burden to them; neither were they intoxicated by luxury; nor did wealth deprive them of their self-control; but they were sober, and saw clearly that all these goods are increased by virtuous friendship with one another, and that by excessive zeal for them, and honour of them, the good of them is lost, and friendship perishes with them.

"By such reflections, and by the continuance in them of a divine nature, all that which we have described waxed and increased in them; but when this divine portion began to fade away in them, and became diluted too often, and with too much of the mortal admixture, and the human nature got the upper-hand, then, they being unable to bear their fortune, became unseemly, and to him who had an eye to see, they began to appear base, and had lost the fairest of their precious gifts; but to those who had no eye to see the true happiness, they still appeared glorious and blessed at the very time when they were filled with unrighteous avarice and power. Zeus, the god of gods, who rules with law, and is able to see into such things, perceiving that an honourable race was in a most wretched state, and wanting to inflict punishment on them, that they might be chastened and improved, collected all the gods into his most holy habitation, which, being placed in the centre of the world, sees all things that partake of generation. And when he had called them together he spake as follows:"

[Here Plato's story abruptly ends.]

APPENDIX C.

IN a pamphlet, in reply to an interesting article in the *Month*, by the Rev. H. W. Lucas, in exposition of what he happily termed " the nature myth " theory, I ventured to suggest a counter-theory for the prominence of the bull in connection with the Diluvian tradition; and as I am not aware that it has been advanced before, and as I have referred to it in the text, I reprint it here.

One of the myths which Mr. Lucas brings in illustration is the myth of Indra, and as he is a personage of the Rigveda, and has good claims to be regarded as primitive, Mr. Lucas is fully justified in doing so. Mr. Lucas says, "There can be no doubt that in the Vedic hymns Indra is the rain-giving firmament." With Mr. Cox he is "the" solar god; " the god of the bright heaven;" " one of the powers which produce the sights of the changing sky," and also "the giver of rain," and " the rain-*bringer* " (Cox, *Aryan Myths*, i. 336-41). Signor Angelo de Gubernatis, *Zool. Myths*, i. 8, says somewhat obscurely, " Like the winds, his companions, the Sun Indra, the sun (and the luminous sky) hidden in the dark, who strives to dissipate the shadows—the sun hidden in the clouds, that thunders and lightens to dissolve it in rain, is represented as a *powerful bull*, as the *bull of bulls*, invincible son of the cow, that bellows like the Maratas." If the mythologists will forgive me, I will endeavour to take this bull by the horns. Mr. Cox, I observe, keeps this name of the Bull in the background when speaking directly of Indra, although when speaking of bulls in mythology he refers to it.

But how came the luminous sky to get these incongruous associations with the bull? All are agreed that it is through Indra's connection with rain. Still, why the identification with the bull, "the bull of bulls"? Signor Angelo de Gubernatis comes to the rescue and says, "To increase the number of cows . . . . is the dream—the ideal—of the ancient Aryan. The bull, the fœcundator, is the type of male perfection, the symbol of regal strength. *Hence* it is only natural that the two prominent animal figures in the mythical heaven should be the cow and the bull." This reasoning may appear very cogent to Signor Angelo de Gubernatis, and "may be highly creditable to him," but for man in his infancy, man in the gelatinous stage of pure imagination, it must have been a great effort, and an effort to which the poetical faculty could have contributed little assistance. It amounts to this, that as according to primitive observation grass grew when the rain fell, and that when the grass grew more bulls would fatten, and that when the rain fell it fell from heaven, therefore the heaven above must be a great cow or bull, *or contain* a great cow or bull! It is easy to see, however, that this anomalous introduction of the bull ill accords with Mr. Cox's really poetical and refined speculations; and, accordingly, by a transcendent effort of the imagination, he makes the Bull Indra identical with the "Lord of the pure ether"! (i. 437.) Surely a greater incongruity was never conceived by mythologist or poet.

Let us now see how this question can be worked on the lines of tradition.

Mr. Cox (i. 336) admits that the myth of Indra may embody a religious idea: "that a moral or spiritual element may be discerned in some of the characteristics of this deity is beyond question, that the whole idea of the god can be traced to the religious instinct of mankind, the boldest champion will scarcely venture to affirm." Neither is it necessary to affirm it. The admixture of the mythic and historical is no difficulty for us, but it is fatal to Mr. Max Müller's theory, unless the religious or historical element can be proved to be secondary.

Now the counter traditional explanation which I shall offer must be regarded as primary and fundamental, as it is intimately connected with the Deluge, and goes back to the second commencement of the human race. There is a tradition that Noah entered the ark when the sun was in the sign of the Bull in the Zodiac.

It will hardly be disputed that this was a date which would have impressed itself upon the recollection of mankind. It was the date when the unintermittent rain commenced, and the sign of the Bull therefore would naturally henceforward have been associated in men's minds with rain and water, whether in fear remembered in connection with havoc and destruction, or whether in pleasant anticipation as the catastrophe faded from recollection, and in the Indian plains drought came to be the greater evil deprecated.

In all the Diluvian traditions the alternations from gloom to rejoicing, the transitions from destruction to subsequent renovation, the contrasts of death and life, have been as much remarked, and are as much in keeping with the narrative in Genesis, as are the contrasts of light and darkness with the solar myths.

Let us now listen to some of those Vedic songs, and see if this key does not equally "unlock the Mythology." Mr. Lucas's quotation, from the Rigveda, shall have precedence (the *Month*, p. 192).* "I declare the *former* valorous deeds of Indra [the Bull] which the thunderer has achieved, he *clove the cloud:* he *cast the waters down*, he broke (a way) for the torrents of the mountains. . . . . *Inasmuch*, Indra, as thou hast struck the first-born of the clouds [or first struck the clouds poetically with the point of his horns, commencing the Duluge] thou hast destroyed the *delusions of the deluders*, and thus engendering the sun, the dawn, and the firmament, thou hast not left an enemy to oppose thee. . . ." Everything here is in keeping except the engendering the Firmament. Taking their statement it would result, however, that the Bull or Indra could not be identical with the firmament, as in some sense of their own he was supposed to *precede* it or *engender* it. But if the firmament meant was the new firmament which arose after the Deluge after forty days and forty nights' rain, the notion might be connected with reminiscences of renovation after the Deluge, when Indra, " as the sun," " in the serene heaven shone out when the deluging clouds had passed away " (Cox, i. p. 337).

Let us now pursue the career of Indra, in the pages of Mr. Cox, "although he has" (p. 339) "but little of a purely moral or spiritual element in his character. . . " "It is true that he is sometimes invoked as witnessing all the *deeds of men*, and thus as taking *cognizance of their sins*." Of Indra, at p. 340, it is said, " Thou thunderer hast shattered with thy bolt the broad and massive cloud into fragments, and hast sent down the

* Mr. Charles E. Govat (*Journal of the Roy. Asia. Soc.*, new series, vol. i. p. 1, 1870), in his account of "the Pongol festival in Southern India," says, " Krishna is always declared by the Brahmins to be the Pongol god ; but the tradition itself bears witness that the *feast is older* than the god. The tale is that when the great wave of Krishna-worship passed over the peninsula, the people were so enamoured of him that they *ceased* to perform the Pongol rites *to Indra*. This made the latter deity *so angry that he poured down a flood upon the earth*."

waters that were confined in it to *flow at will;* verily thou
alone possessest all power." "At the birth of thee who
art resplendent ["resplendent" would apply to a star or
constellation, and by its birth might be intended the new·
era, which was inaugurated by its appearance or prominence in the heavens at the commencement of the Deluge]
trembled the earth through fear of thy *wrath,* the mighty
clouds were confined; they *destroyed,* spreading the waters
over dry places." Mr. Cox interpolates thus—"destroyed
(the distress of drought);" but that is only a gloss of his
own, and the "dry places" may signify high and dry,
places never reached by the waters before. At p. 342 we
find Indra especially described "as the god of battles, the
giver of victory to his worshippers, the destroyers of the
enemies of *religious rites,* the *subverter* of the *cities* of the
Asuras."

In De Gubernatis (i. p. 9) it is said that the cloud in
the Rigveda is often represented " as an immense great-
bellied barrel (Kabandkas), which is carried by the divine
Bull." "The terrible bull bellows and shows his strength
and sharpens his horns, the splendid bull with sharpened
horns, *who is able of himself to overthrow all peoples.*"
"The bull Indra is called the bull with the thousand
horns who rises *from the sea.*" But these verses were
composed by a pastoral people, who, if the descendants of
Japhet, had probably always dwelt inland since the Deluge. At p. 19, Indra is represented as discomfiting "the
monster (rakshohanân) who destroys by fire the monsters
that live in darkness." Another Vedic hymn informs us
that the monster Valas had the shape of a cow; another
hymn represents the cloud as the cow that forms the
waters, and that has now one foot, now four, now eight,
now nine [more applicable to the constellation than to the
rain-giving firmament], and fills the highest heavens with
sounds [commencement of the Deluge]; still another

hymn sings that *the sun hurls* his golden disc *in* the variegated cow ; they who have been carried off, who are guarded by the monster serpent, the waters, the cows become the wives of the demons."

This will suffice to show that these poems may be read in another than the present popular sense. It must be noted that the cavern in which the cows are concealed is also, according to Bryant and Faber, a common form of the tradition of the ark. Moreover, " three or seven brothers and sisters figure in these conflicts and adventures. The number three corresponds with the sons of Noah, the number seven with the persons saved in the ark, if referred to separately from the Patriarch. As to the recurrence of the numbers seven and eight, *vide Tradition*, p. 198 : compare the representation of Horus, who " is frequently represented as the eighth, conducting the *bark* of the gods with the seven great gods." The Patriarch might get detached in tradition in several ways, *e.g.* when he is located in or identified with the sun, and the rest in other parts of the heaven. In the Vedas, however, when mentioned in connection with the Bull, " the seven shiners " may very well be the *seven* brilliant stars, which form the Pleiades in the constellation Taurus (in the neck of the Bull). The Hyades in the same constellation, Mr. Lucas tells us, were associated with " moisture," and it might have been added, as the " tristes Hyades," with ill-omen to mariners.

It may be asked whether, beyond the indirect, I have any direct testimony to this tradition, that Noah entered the ark when the sun was in the sign of Taurus or the Bull. I cannot recall at this moment where I met with the tradition in the first instance, but I have since come upon confirmatory evidence, with which I shall conclude, as I think it will sufficiently establish my point.

As the Zodiac commences with Aries, it is presumable

that the primitive months commenced their series also with Aries, the Ram.

The traditional character of the Zodiac has often been remarked—and this is fully confirmed by Mr. George Smith's recent Assyrian discoveries. In his *Chaldæan Account of Genesis*, p. 69 :

" In the fifth tablet of the *Creation* legend we read:

1. ' It was delightful all that was fixed by the great gods.

2. ' Stars, their appearance [in figures] of animals he arranged.

3 ' To fix the year through the observation of their constellations

4. ' Twelve months (or signs) of stars in three rows he arranged.' "

And at p. 73, Mr. G. Smith says :

" We then come to the *creation* of the heavenly orbs, which are described in the inscription as arranged *like animals*, while the Bible says they were set as ' lights in the firmament of heaven ;' and just as the book of Genesis says they were set for signs and seasons, for days and years, so the inscription describes that the stars were set in courses to point out the year. The twelve constellations or signs of the Zodiac and two other bands of constellations are mentioned, just as two sets of twelve stars each are mentioned by the Greeks, one north and one south of the Zodiac."

In our tables, Taurus, or the bull, is the *second* sign of the Zodiac. Now Genesis (vii. 11) tells us that Noah entered the ark on the 17th of the *second* month. Again, I find in Mon. F. Lenormant's *Fragmens Cosmogoniques de Bérose*, 1871, p. 211, that the *second* Assyrian month was named " the Bull," also the Accadian second month, and also the Hebrew. " Les nom des mois juifs sont les noms des mois CHALDEO-assyriens."

I see further trace of the tradition in the following passage from Canon Rawlinson's *Illustrations of the Old Testament*, p. 18, bearing in mind that the Vedic legends of the dog (*canis major*) are connected with the legends of Indra (De Gubernatis, ii. 19). " The Cherokee Indians had a legend of the destruction of mankind by a deluge, and of the preservation of a single family in a boat, to the construction of which they had been *incited by a dog*."

Plutarch, in his treatise *De facie in orbe Lunæ*, c. 26 (*vide* Sir G. Cornewall Lewis's *Astronomy of the Ancients*, p. 491), says that the people of the island of Oxygia (compare deluge of Oxyges*) "pay the principal honours to Saturn, and after him to Hercules. When the planet *Saturn* (compare analogy between Saturn and Noah, *Tradition*, &c., p. 211) *is in* the sign of *Taurus* (the bull), a coincidence which occurs every thirty years, they send out a body of men, selected by lot, *to seek their fortunes across the sea*." Mr. Lucas's instinct as a mythologist will tell him that true tradition may be found embodied in fabulous narration, and he will not peremptorily reject the evidence, even if the geographical latitude of Oxygia is not precisely ascertainable.

These instances are taken more or less casually, and the list is far from being exhausted. It is very probable that other traces might be found. However widely we may differ, we are agreed as to the importance of the inquiry. There are those who deride the study of Mythology, although we might truly extend the dictum of Proudhon, that "au fond de toute question on trouve toujours la théologie," and add, "et la Mythologie." The prominence given to it in current literature is in attestation of this remark.

No theory, not excepting Darwinism, so seductively

---

* "The Greeks had two different traditions as to the Deluge which destroyed the primitive race. The first was connected with the name of Oxyges . . . *personnage tout à fait mythique*. . . . His name even was de·ived from a root or word (*de celui qui*) which originally imp ied the Deluge (in Sanscrit 'Augha'). They narrated that in his time the whole country was invaded by the Deluge, the waters of which reached the heavens, and from which he escaped in a vessel, with some companions. The second tradi·ion was the Thessalian legend of Deucalion" (F. Lenormant, *Manuel d'Hist. Ancienne*, ii. p. 69).

takes back the history of the human race, and shrouds it in the indefinite lapse of ages, as the mythology of the winds and the elements.

Mine is perhaps only a feeble attempt to get, so to speak, at the back of the North Wind ; but any attempt which succeeds in doing so will reveal as much to science (however grim the revelation and disappointing to golden dreams) as the discovery of the North Pole.